T0313227

PHYSICS OF
Nuclear Radiations
Concepts, Techniques and Applications

PHYSICS OF
Nuclear Radiations
Concepts, Techniques and Applications

Chary Rangacharyulu

CRC Press
Taylor & Francis Group
Boca Raton London New York

CRC Press is an imprint of the
Taylor & Francis Group, an **informa** business

Taylor & Francis
Taylor & Francis Group
6000 Broken Sound Parkway NW, Suite 300
Boca Raton, FL 33487-2742

© 2014 by Taylor & Francis Group, LLC
Taylor & Francis is an Informa business

No claim to original U.S. Government works

Printed on acid-free paper
Version Date: 20131106

International Standard Book Number-13: 978-1-4398-5777-9 (Hardback)

Visit the Taylor & Francis Web site at
http://www.taylorandfrancis.com

and the CRC Press Web site at
http://www.crcpress.com

Contents

Preface

Most pro-nukes vouch for the safety of nuclear radiations and say not to be concerned about possible hazards. Anti-nukes, naturally, take the opposing position and proclaim that exposure to nuclear radiations, however small the dose may be, is harmful. The population, both well educated and not so well educated, take a position mostly based on how well the proponents articulate their viewpoint. People who vehemently oppose nuclear power reactors or nuclear applications in agriculture do not express strong opposition to medical installations, either particle accelerators or administration of nuclear radiations for medical diagnostic or therapeutic purposes.

I, having worked with nuclear radiations of various species and a wide range of energy at several nuclear installations in North America, Europe and Asia along with junior and senior researchers for over 40 years, formed the view that the physics of nuclear radiations can be made accessible to a wider community of students and professionals with some basic background in physics and mathematics.

With this view, I designed and started offering a course on physics concepts of nuclear radiations about 5 years ago at the University of Saskatchewan. This course evolved, but the basic requirement is that a student has first year university mathematics and physics knowledge. I do not get into convoluted mathematical derivations, but present the mathematical formulae and discuss their meaning and the domains of applicability of the equations at hand. I remind students that mathematical symbols and equations are simply vocabulary and phrases, which the physics community makes extensive use of. The purpose of the course is not to convince students one way or the other about the hazards of nuclear radiations. It is to empower them with tools so they can calculate and assess nuclear radiations and their impact.

The student population was quite diverse, composed of physics majors, engineers and health science and education students. During the summers of 2012 and 2013, I offered an intensive short course for health professionals and graduate students of East Africa at the Nelson Mandela African Institute of Science and Technology, Arusha, Tanzania. In all these situations, the students are better informed about nuclear radiations at the end. That is, at least, the feedback I have gotten so far.

This book is an outgrowth of these courses. Here we take a simple approach. The first chapter is a review of physics preliminaries, of energy and momentum conservations, etc., and it provides students with rules of thumb and fingertip information, to set them thinking along these lines. The first consideration for any nuclear radiations should be how long it will last as it is being emitted by nuclear decays or activation. This is dealt with in Chapter 2.

The next question is what is the stability of nuclei or particles against radioactive transformations. We may be interested in decay processes or the reactions that we induce at particle accelerators or nuclear reactors. The question is whether a hypothetical transformation emitting hazardous radiations or an alchemy to make a profit can happen. The first and foremost consideration is the energy balance. This topic is dealt with thoroughly in Chapter 3.

Once radiations are emitted, the next question is how far they travel in a material medium and what are the energy deposits and ionizations along their paths. In this regard, heavy charged particles behave quite distinctly from electrons, photons and neutrons. Conceptually, charged particle behavior is very predictable, as they lose energy continually. This is the topic of discussion in Chapter 4.

We then take up electrons and photons as one topic due to the inseparability of high energy electrons and photons as far as the energy or intensity loss processes are concerned. Electrons emit radiations and photons create electron-positron pairs and their propagation is very much intertwined. Chapter 5 presents these details.

The energy loss of charged particles or attenuation of photons is mainly through electromagnetic interactions. On the other hand, neutron attenuations are dominantly through nuclear interactions. They do not show any monotonous variations with changing mass number or atomic number. Chapter 6 address these interactions.

Chapter 7, then, addresses dosimetry. Dosimetry is a multidisciplinary subject which encompasses physics at first and then involves biology, physiology, etc. As a result, dosimetry is more an art than a hardcore science. We limit our discussion in this subject to nomenclature and physics reasoning and refrain from entering into a terra ambiguus.

In addition to nuclear power and research reactors, several nuclear facilities, both small and large, are radiation sources. Chapter 8 is a discussion of typical facilities such as medical X-ray machines and particle accelerators. The emphasis is to recognize that physics principles are simple and straightforward. It is the ingenuous application and extension of physics principles which brings the development of advanced radiation facilities. At the heart of any nuclear radiation measurements are radiation detectors. It is important to realize that detectors of all types rely on diverse interactions of radiation with matter and information transport to external signal recognition and processing systems, either electronic or other types. The physics principles of diverse detectors are discussed in Chapter 9.

Nuclear measurements range from a single detector to highly complex systems of detector assemblies, all working in concert to generate the data of interest for safety or science or other applications. But, at the basic level, they all function on simple basic physics and instrumentation logic. It is only the multiplicity of a system and the diversity of radiations which change from a simple one to a highly sophisticated measurement assembly. These points are emphasized in Chapter 10, which illustrates methods of energy and time spectroscopies.

Interspersed throughout the text are applications in archeology and science. It is assumed that everyone knows about nuclear power. Thus, Chapter 11 highlights a few applications in agriculture, medicine, industry and one example of art.

My sincere thanks are due to Cody Crewson for his expert LaTeX typesetting and formatting. He has put up with me for changing drafts and last minute demands on his time and expertise. He has been extremely patient with me.

Francesca McGowan of Taylor & Francis persisted that I finish the drafts. She has been very gentle but certainly very persuasive to see this book done. I owe her a big thank you.

1

Physics Preliminaries

The study of nuclear radiations, like most other physics topics, requires that we have some basic physics vocabulary at our fingertips. The vocabulary and phrases are expressed as mathematical equations connecting the physical properties of entities such as mass, electric charge, energy, momentum, wavelength, frequency, etc. Almost all estimates of radiations involve understanding its inter-relationship and being able to manipulate the equations. As we manipulate them, we should always bear in mind their range of applicability. This chapter will provide you with the necessary equations. We will introduce the units of these physical parameters as they are routinely used in radiation physics calculations.

1.1 Conservation Laws

Conservation principles play an essential role in all physical phenomena. In model descriptions of nuclear radiations, we rely on several physical and abstract conservation principles. The energy, momentum, angular momentum and electric charge conservation principles are of primary interest.

We assert that the total energy of a system is conserved in all physical processes. We start off with an arbitrary number of entities (photons, particles, etc.) in an initial state that evolves to a final state. In any progression, at each stage, conservation principles require that the following quantities of the system remain constant.

Conserved quantities In physics, the classical conservation principles of total energy, linear momentum and electric charge are always satisfied. So far, no phenomenon that disobeys any one of these laws has been discovered. These three conservation laws are discussed below.

Conservation of Total Energy Mass is a form of energy. Kinetic energy and mass are attributes of a body in motion. Kinetic energy and mass are separately conserved in elastic processes. The departure from classical concepts is that mass is not necessarily conserved in atomic or nuclear processes. Inelastic processes, involving emission or absorption of radiation, or changing the composition of interacting partners, do not conserve the mass or kinetic energy. Total energy, however, is conserved.

Prior to relativity, inelastic processes accompanied by energy dissipation were known. This is what happens in chemical reactions resulting in the formation of compounds. In relativity, we realize that processes where interacting partners reorganize themselves to another set of constituents can result in non-conservation of total mass. Also the kinetic energy of the system is not conserved in inelastic processes. The problem is now simplified since we mainly require the conservation of two kinematic parameters, i.e., total energy and total momentum. It is interesting to note that such transformations almost always lead to loss of total mass, consistent with the principle: left to itself, a system evolves to a state of *lowest energy*. As we will see in Chapter 2, nuclear decay processes are governed by this principle.

The Conservation of Linear Momentum This physics principle is as old, at least, as Newton's Laws of Motion. It still holds very rigorously for all processes. Linear momentum is a vector quantity. We require that the vector sum of momenta of all interacting partners remain unchanged during the process. This implies that the momentum components along each of the mutually orthogonal directions must be separately conserved. This requirement has important implications for the directions of the motion of particles in the laboratory. For example, when a particle at rest emits two particles, the products will move in opposite directions. They are 180° apart. By measuring the momentum and direction of one particle, we can assert the corresponding parameters of the partner radiation.

Conservation of total energy and linear momentum are sufficient to perform quantitative analyses of collisions and decay processes.

Conservation of Electric Charge The total amount of electric charge is conserved. However, this principle allows for creating equal amounts of positive and negative charges in particle decays or a reaction and still satisfying the charge conservation. It is a very important distinction from other conservation principles. The energy conservation principle makes it possible to transform one form of energy into another form, but we do not create energy in the process. In high energy collisions and photon interactions, copious amounts of particles and antiparticles of opposite charges are produced and annihilated. Thus, at any stage of the reaction or decay, there may be an arbitrary number of positive and negative charges flying about the laboratory as long as the net charge remains constant.

We summarize the above conservation principles below:

Energy (E) conservation (a **scalar quantity**)[1]:

$$\sum E_{initial} = \sum E_{final} \qquad (1.1)$$

$$\text{or}$$

$$\Delta E = 0$$

Momentum ($\vec{p}$) conservation (a **vector quantity**):

$$\sum \vec{p}_{initial} = \sum \vec{p}_{final} \qquad (1.2)$$

$$\text{or}$$

$$\Delta p_i = 0$$

where

$$i = x, y, z \text{ in cartesian coordinates}$$
$$i = r, \theta, \phi \text{ in spherical polar coordinates}$$

It should be remarked that arbitrary amounts of momentum can be carried by particles as long as the following conditions are satisfied:

---

[1] Verifying the conservation principle of a scalar quantity is simple addition and subtraction of numbers. For a vector quantity, the conservation of components along the three mutually perpendicular (orthogonal) directions is essential.

a) The components along three orthogonal directions are conserved. Here, what matters is the overall algebraic sums of momentum in each direction and not the momentum of individual entities separately.

b) The total energy conservation principle is obeyed.

A familiar example is the radioactive decay of an atomic nucleus. Initially, the nucleus is at rest with zero momentum. It emits two particles, say an alpha particle and another nucleus. They fly in opposite directions, with each carrying same magnitude of momentum. Each particle has a finite momentum and the vector sum of the momenta is zero.

Electric Charge (Q) conservation (a **scalar quantity**):

$$\sum Q_{initial} \;=\; \sum Q_{final} \qquad (1.3)$$

or

$$\Delta Q_{positive} \;=\; \Delta Q_{negative}$$

The last equation is simply a mathematical expression that equal amounts of positive and negative charges may be created or annihilated in nuclear processes.

1.2 Basics of Relativity

Among the most commonly encountered species of radiations are X-rays or gamma rays of zero mass with the collective name of photons,[2] traveling at the speed of light c. Newtonian relativity is not directly applicable and we must resort to kinematic equations of special relativity.[3]

---

[2]Experimental works so far find the mass of a photon $m_\gamma < 10^{-17}$ eV according to The Particle Data Group's Review of Particle Physics, Journal of Physics G, Volume 37, Number 7A, July 2010. In everyday units, photon mass is less than 10^{-53} kg. In a vacuum, photons travel at speed $c = 3 \times 10^8$ m/s.

[3]There are several books and I am sure that there are innumerable websites which provide a thorough introduction to the physics reasonings and mathematical formulations of special relativity of interest to us. A classic book on this subject is Special Relativity by A.P. French, published by Taylor & Francis (1968).

In special relativity, the mass of an object is a form of energy and it varies as the velocity of an object changes. At first, this might sound strange but, upon a little reflection, it is not that bizarre if we stick to the definition that mass is a measure of inertia.

From experience, we know that the influence of a force on a moving body is different from that on the same body at rest. We may quantify this effect as a change in the inertia of a body as it moves.

However, we must assign a specific mass to a particle or entity which would unambiguously identify what we are referring to. In physics, we assign mass to each particle, nucleus, atom or molecule that would be measured by an observer[4] who is at rest relative to the object. This is called rest mass.

Rest mass is the smallest mass that we measure for a particle. Unless otherwise specified, the masses we refer to in this book are the rest masses of particles or nuclei.

For a particle moving at speed v, with a momentum p, the total energy E is the sum of the translational energy T and the energy due to its rest mass ($m_0 c^2$). Special relativity gives relations between mass (m), velocity (v), momentum (p), kinetic energy (T) and energy (E) as:

$$\vec{p} = m\vec{v} = \gamma m_0 \vec{v} \tag{1.4}$$
$$E = T + m_0 c^2 \tag{1.5}$$
$$E^2 = p^2 c^2 + m_0^2 c^4 \tag{1.6}$$

Here, m_0 is mass expressed in grams or equivalently in units[5] of electron volts per c^2 [eV/c^2]. The speed of light in a vacuum, c, is a fundamental constant in physics. It is interesting to note that the momentum $\vec{p} = m\vec{v}$ is exactly in the form we are used to in Newtonian mechanics, i.e., momentum = mass × velocity. However, here the mass m is not a constant, as it depends on the object's speed and it is related

---

[4]We use the term observer to mean a person or a measuring instrument involved in observation.

[5]See Section 1.7.

to the constant rest mass (m_0) by simple relations as follows:

$$m = \gamma m_0 \tag{1.7}$$

$$\gamma = \frac{1}{\sqrt{1 - \beta^2}} \tag{1.8}$$

$$\beta = \frac{v}{c} \tag{1.9}$$

In the above equations β is the ratio of the speed of the object to the speed of light in a vacuum. Thus, β and the Lorentz factor γ are dimensionless numbers.

It is useful to realize that $0 \leq \beta \leq 1$ and $\gamma \geq 1$, since no material object with a finite mass ($m_0 \neq 0$) can be accelerated to the speed of light c. Mathematically, we notice that for $\beta = 1$, mass is infinity for finite rest mass and for $\beta > 1$, γ is an imaginary number. Thus $\beta > 1$ is not meaningful for physics discussions.[6] Observers moving at different velocities with respect to the object deduce different magnitudes of β and γ. Each observer infers a different mass for the same object.

1.3   Energetics of Photons

Photons, which may be called electromagnetic quanta,[7] are unique in their properties. The energy E is intrinsic to these quanta of zero mass. They can be neither accelerated nor decelerated. Even though it may be tempting to identify their energy with translational energy, it is not necessary to be concerned about its exact nature. For all practical purposes, it suffices to note that for photons

$$E = h\nu = \frac{hc}{\lambda} \tag{1.10}$$

---

[6] You might have heard about particles moving faster than the speed of light. They are called tachyons. They are yet to be discovered and we do not come across those objects in our discussions. Recent excitement was about neutrinos moving at speeds faster than that of light. The wrong experimental result was due to a faulty cable. Physicists are not infallible after all.

[7] The physical nature of the photon is not yet understood. It plays the role of information carrier or a messenger in electromagnetic processes. Physicists call photons electromagnetic quanta to avoid using the term particles for them.

where h is Planck's constant,[8] a fundamental quantity in physics. The frequency ν and wavelength λ of photons are related as

$$\nu \times \lambda = c \equiv 3 \times 10^8 \text{m/s} \tag{1.11}$$

Depending on the application at hand, we may want to identify radiation by its wavelength, frequency or energy. It is useful to recognize the proportionality relations among these variables to make easy conversion from one to the other.[9]

From Equation 1.10 we note that the product of energy times wavelength is hc, a product of two fundamental constants and thus itself a constant. By straightforward multiplication,

$$E\lambda = hc \tag{1.12}$$
$$= 4.135 \times 10^{-15} \, [\text{eV} \cdot \text{s}] \tag{1.13}$$
$$\times 3 \times 10^8 \left[\frac{\text{m}}{\text{s}}\right]$$
$$= 1.24 \times 10^{-6} \, [\text{eV} \cdot \text{m}]$$
$$= 1240 \, [\text{eV nm}] \tag{1.14}$$

If a photon's energy is expressed in electron volts, its wavelength is

$$\lambda = \frac{1240}{E \, [\text{eV}]} \, [\text{nm}] \tag{1.15}$$

It is easy to recognize that photon energies in multiples of eV are related to wavelengths in sub-multiples of wavelength by this constant.

$$E \times \lambda = hc = 1240 \begin{bmatrix} \text{eV} \cdot \text{nm} \\ \text{keV} \cdot \text{pm} \\ \text{MeV} \cdot \text{F} \\ \text{GeV} \cdot \text{am} \end{bmatrix} \tag{1.16}$$

Thus, a photon of 1 eV energy would correspond to a 1240 nm wavelength (infrared), while that of 1 keV energy is 1240 pm (or 1.24

[8]See Section 1.7.
[9]See Section 1.7.

nm, an X-ray), etc. This observation has important practical implications. From optics, we know that the wavelength of a probe must be comparable to the sizes of objects and apertures being investigated. As we probe smaller and smaller sizes, we must employ radiations of decreasing wavelengths or equivalently increasing energy for photons.

For crystal spacings and atomic dimensions of about a nanometer, the corresponding energies are a few keV (soft X-ray region); for nuclear dimensions of a few F (Fermi or femto meters, $F = 10^{-15}$ m), photons are of several MeV. In particle physics, as we deal with photons of a few GeV, we probe distances of about 10^{-18} m. Corresponding frequencies can be readily calculated.

From Equation 1.6, we note that photon momentum ($m_0 = 0$) can be written as

$$p = \frac{E}{c} = \frac{h\nu}{c} = \frac{h}{\lambda}. \tag{1.17}$$

It is useful to note that momentum can be written in units of energy per c such as [keV/c], [MeV/c], etc. Thus, we can write kinematic parameters in units of E/c^x, with $x = 0$, 1 and 2 for energy, momentum and mass, respectively. The compactness and convenience of this notation comes in handy as we compare one physical parameter (mass, momentum, energy) to another one.

1.4   Energetics of Matter

Any material object of finite rest mass ($m_0 \neq 0$) is limited to speeds $v < c$ or $\beta < 1$. From the well known relation (Eq. 1.6),

$$E^2 = p^2c^2 + m_0^2c^4 \tag{1.18}$$

we write

$$\begin{aligned}
E^2 &= m^2v^2c^2 + m_0^2c^4 \\
&= m_0^2c^4\left[\gamma^2\beta^2 + 1\right] \\
&= \gamma^2 m_0^2 c^4 \\
&= m^2c^4
\end{aligned} \tag{1.19}$$

Thus we arrive at the most celebrated Einstein equation

$$E = mc^2 = \gamma m_0 c^2 \tag{1.20}$$

From this equation, we see that

$$m = \frac{E}{c^2} \tag{1.21}$$

or mass can be written in units of energy/c^2.

As energy is expressed in units of electron volts, it is convenient to refer to mass in units of energy/c^2, i.e., eV/c^2.

As c^2 is a fundamental constant, physicists use a shorthand notation to give names in units of electron volts.

Example 1.1

Mass of an electron: $m_e = 9.09 \times 10^{-31}$ kg

$$E = m_e c^2$$

$$= 9.09 \times 10^{-31} \times \left(3 \times 10^8\right)^2$$

$$= 81.8 \times 10^{-15} \text{ J}$$

or mass of an electron

$$m_e = \frac{E}{c^2} = 81.8 \, \frac{\text{J}}{c^2}$$

$$1 \text{ eV} = 1.602 \times 10^{-19} \text{ J}$$

$$m_e = 81.8 \times 10^{-15} \, [\,\text{J}\,] \times \frac{1}{1.602 \times 10^{-19}} \left[\frac{\text{eV}}{\text{J}}\right] \times \frac{1}{c^2}$$

$$= 0.511 \times 10^6 \left[\frac{\text{eV}}{c^2}\right]$$

$$= 0.511 \left[\frac{\text{MeV}}{c^2}\right]$$

Thus, the mass of an electron is 0.511 MeV/c^2 in energy units. The conciseness of the unit, instead of kilograms, is clear. Practitioners, in general communications, drop the mention of c^2, without losing the significance. If we say that an electron has a mass of 0.511 MeV, it is understood as 0.511 MeV/c^2.

A few more kinematic relations are of interest. Kinetic energy is given by

$$T = E - m_0 c^2 \tag{1.22}$$

or

$$T = (\gamma - 1) m_0 c^2 \tag{1.23}$$

Note that this equation is different from Newton's equation of kinetic energy

$$T = \frac{1}{2} m_0 v^2 = \frac{1}{2} m_0 \beta^2 c^2 \tag{1.24}$$

At low speeds $v \ll c$ or $\beta \ll 1$, Newton's equation for kinetic energy is a good approximation for Einstein's equation. While Einstein's equations are valid for all energies and at all speeds, Newton's equations are applicable at low speeds only. When in doubt, the use of relativistic expressions is strongly recommended.

Numerical values of kinematic parameters of interest can be easily calculated if we become familiar with some simple recipes of manipulating algebraic relations.

For momentum, we have

$$pc = \sqrt{E^2 - m_0^2 c^4}$$

$$= E \sqrt{1 - \frac{1}{\gamma^2}} \tag{1.25}$$

$$= E\beta$$

$$\text{or} \quad p = \frac{E\beta}{c} \tag{1.26}$$

It is of interest to note that the expressions of momentum for material particles and photons differ only in that there is an extra dimensionless factor β for material particles. For photons $\beta = 1$, $m_0 = 0$ and $E = h\nu$. From this observation, we note that the kinematic expressions for photons and material particles are not dissimilar, after all. At the same time, it is worth noting that a material body cannot attain the speed of light, while a photon cannot be slowed down.

As stated above, momentum is commonly expressed in units of E/c. It is especially useful for very relativistic particles ($E \sim T \gg m_0c^2, \beta \sim 1$), for which energy and momentum are of nearly the same numerical value when expressed in these units.

Example 1.2

The rest mass of an electron is $m_e = 9 \times 10^{-31}$ kg $= 0.511$ MeV/c^2. When accelerated through a potential difference of 100 million volts, it gains a kinetic energy of 100 MeV. Then, its total energy is $E = T + m_ec^2 = 100.511$ MeV. Its relativistic mass is $m = E/c^2 = 100.511$ MeV/c^2. The relativistic factor (known as the Lorentz factor) is

$$\gamma = E/m_ec^2 = \frac{100.511}{0.511} = 199.7$$

$$\beta = \frac{v}{c} = \sqrt{1 - \frac{1}{\gamma^2}} = \sqrt{1 - \frac{1}{1997.2^2}} = 0.99995$$

$v = 0.99995 \times c = 99.995\%$ of the speed of light.

The momentum is $p = E\beta/c = 100.511 \times .99995/c = 100.5$ MeV/c. If we use the Newtonian equation, we get

$$T = \frac{1}{2}m_0v^2 = \frac{1}{2}m_0\beta^2c^2 = \frac{1}{2} \times 0.511 \times \beta^2 = 100 \text{ MeV}$$

The velocity of an electron turns out to be about 20 times the speed of light, inconsistent with the prescription of special relativity.

Instead of an electron, let us consider a proton of mass $m_p = 938.3$ MeV/c^2, accelerated to gain a kinetic energy $T = 100$ MeV. Total energy $E = 938.3 + 100 = 1038.3$ MeV. The Lorentz factor is $\gamma = E/m_pc^2 = 1038.3/938.3 = 1.1066$.

$$\beta = \frac{v}{c} = \sqrt{1 - \frac{1}{\gamma^2}} = \sqrt{1 - \frac{1}{1.1066^2}} = 0.428$$

The speed of a proton is $v = 0.428c$, i.e., it is 42.8% of the speed of light or

$$v = 0.428 \times 3 \times 10^8 \text{ m/s} = 1.285 \times 10^8 \text{ m/s}$$

If we apply Newton's equation,

$$T = \frac{1}{2}m_0v^2 = \frac{1}{2}m_0\beta^2c^2 = \frac{1}{2} \times 0.938.3 \times \beta^2 = 100 \text{ MeV}$$

$v = 0.462c$ or it is 46.2% of the speed of light. Newton's expression still overestimates the speed of a particle, albeit by a small amount of about 4%. In such a case, Newton's expression is a reasonable approximation unless the experimental conditions demand higher precision. However, it must be remembered that special relativity gives the correct result for all speeds.

Many experiments involving charged particles entail bending them in magnetic fields. It may be an accelerator system bending them to direct charged particles to a target point in the laboratory or a mass spectrometer for research, pharmacokinetics of blood or urine analysis or protein characteristics. It may be we are interested in knowing the magnetic field strength required to affect a bending path for a particle to follow or one would like to know the momentum of a particle as it passes along a path in a known magnetic field. In either case, a simple formula is all that is needed:

$$p = 0.3 \times B \times R \qquad (1.27)$$

In equation 1.27, p is the momentum of a particle of unit charge e (one electronic charge), B is a magnetic field perpendicular to the particle direction and R is the radius of curvature of the particle in the field. Here, the momentum is in units of [GeV/c], B is in units of Tesla [T] and R is in meters.[10] For fixed magnetic field direction and momentum vector direction, the sign of the radius of curvature gives the sign of the electric charge of a particle. Positively charged particles such as protons, positrons, positive ions, etc. are bent with positive curvature, while electrons, negative ions and other negatively charged particles are of negative curvature. Thus these measurements provide information on momentum and magnitude along with the sign of the charge of the particle, whether it is positive or negative.

It is important to note that this measurement of bending in magnetic fields reveals the momentum of a charged particle but not its identity. From the bending of, say, a positively charged particle with 1 GeV/c, we know that it may be a positron (anti-electron $m_e = 0.511$ MeV/c^2),

---

[10]See Section 1.7.

a pion ($m_\pi = 139.5$ MeV/c^2), a muon ($m_\mu = 105$ MeV/c^2) or a proton ($m_p = 938.3$ MeV/c^2), etc. But, from this measurement alone, we can not say which it is.

We must supplement this information with at least another datum, say the speed of the particle β. If the momentum p and speed β are determined, the mass of the particle is easily calculated as

$$m_0 = \frac{pc\sqrt{1-\beta^2}}{\beta} \tag{1.28}$$

In many radiation experiments, when the identities are unknown, a common technique is to determine the speed by measuring the time it takes for a particle to traverse a known distance. It is just like clocking a racer as he or she is competing in track event. There are a few other techniques which we will discuss in later chapters.

Example 1.3

A proton of mass $m_p = 938.3$ MeV/c^2 is accelerated to a momentum $p = 1$ GeV/c. What is its speed? What is the bending radius of this proton if a magnetic field of 1 telsa is applied perpendicular to its direction of motion?

We calculate the total energy

$$E = \sqrt{p^2c^2 + m_0^2c^4} = \sqrt{1 + .938^2} = 1.371 \text{ GeV}.$$

Then we calculate the speed parameter

$$\beta = \frac{pc}{E} = \frac{1}{1.371} = 0.73$$

or speed $v = 0.73 \times c = 2.19 \times 10^8$ m/s.

The kinetic energy of the proton is

$$T = E - m_pc^2 = 0.433 \text{ GeV}. \tag{1.29}$$

In a 1 tesla magnetic field ($B = 1$ tesla), the radius of curvature is

$$R = \frac{p}{0.3B} = \frac{1}{0.3} = 3.33 \text{ m} \tag{1.30}$$

If an electron of the same momentum (1 GeV/c) traverses the same magnetic field, it will bend with a radius of negative curvature of -3.33 meters.

We calculate the total energy of an electron $E = \sqrt{p^2 c^2 + m_0^2 c^4} = \sqrt{1 + .0.00051^2} \approx 1$ GeV.

The kinetic energy is 1 GeV to 5 significant digits. The calculation for β will give a result $\beta = 0.99999$ to 5 significant digits.

The speed of an electron is $v \sim 0.999999 \times c \approx c \approx 3 \times 10^8$ m/s.

Thus the electron is highly relativistic. For many practical purposes, we may consider $E = 1$ GeV, $p = 1$ GeV/c and $\beta \sim 1$ for these electrons.

It is quite gratifying that the physics principles of measurement of momentum, speed and energy are all that is needed to identify a particle. It is enough to measure two of these physical parameters to unambiguously identify the particles, though the practical experiments can be quite challenging and might necessitate a very elaborate measuring apparatus. It may also be necessary to supplement with other information such as energy loss in a medium to get unambiguous results.

It may be of interest to note that an enormous setup called the AT-LAS detector at the Large Hadron Collider laboratory is mainly concerned with measuring electric charges, momenta, speeds and energies of various particles and radiations which result in the collisions of particle beams. You may want to visit the website http://atlas.ch.

1.5 Matter Waves

It is no exaggeration to say that de Broglie's[11] hypothesis of matter waves was a major breakthrough for 20th century quantum mechanics. This hypothesis prompted Schrödinger and Heisenberg to develop mathematical formulations of modern quantum mechanics which did away with several conceptual, logical difficulties of earlier works.

---

[11]Louis de Broglie (1892–1987).

We may summarize the development as follows: Einstein reasoned that light seems to propagate as waves as evidenced by optical diffraction, interference and polarization phenomena. It behaves like a particle in its exchanges of energy and momentum with materials in the medium as evidenced by photovoltaic effect[12] and blackbody radiation. This was the beginning of the photon concept and the dawn of quantum physics.

One might then ask: If light which we consider as waves exhibits material properties, is it not likely that the entities we normally consider as material objects show wave properties? If the answer is yes, what are the observables and where should one look for them?

In optics, diffraction and interference phenomena are observed when the dimensions of apertures are of the same magnitude as the wavelength of light. In his Ph.D. thesis, de Broglie hypothesized that a moving object has matter waves associated with it. The wavelength is inversely proportional to an object's momentum in exact analogy with photons. Thus, we can make quantitative predictions.

The matter waves associated with each body can be assigned a wavelength, the *de Broglie wavelength*, given by

$$\lambda = \frac{h}{p} = \frac{hc}{pc} = \frac{hc}{\sqrt{E^2 - m^2c^4}} = \frac{hc}{\sqrt{(T+mc^2)^2 - m^2c^4}} \qquad (1.31)$$

For non-relativistic cases ($T \ll mc^2$ and $v \ll c$), it simplifies to

$$\lambda = \frac{hc}{\sqrt{2Tmc^2}} \qquad (1.32)$$

If we express mass m in [MeV/c^2] and kinetic energy T in [MeV] units, then the wavelength is in units of femtometers (10^{-15} m) for $hc = 1240$.

The de Broglie wavelength is a semi-quantitative estimate of dimensions at which quantum effects become important. If sizes or spacings are comparable to or smaller than the de Broglie wavelength of the

---

[12]A simple example of photovoltaic effect is solar cells. Solar cells generate electricity only when the sun shines. Lack of visible light during cloudy days or at night renders solar cells inoperable during those times. From classical electricity and magnetism of wave-like behavior, we could expect solar cells to function on hot summer nights even when it is dark.

probe radiation, one has to resort to quantum mechanics. For very large distances or lengths, a classical physics approach would suffice. For practical purposes, the de Broglie wavelength provides an estimate of energies of different species of particles for probing matter of specific dimensions.

Example 1.4

Nuclear research reactors produce high fluxes of thermal neutrons, extensively used for materials research and also for non-destructive testing in industrial applications.

In thermodynamics, the mean kinetic energy of a gaseous particle is given by

$$E_{kin} = k_B T \, [\text{eV}] \tag{1.33}$$

where the Boltzmann constant k_B is 1.38×10^{-23} [J·K^{-1}] or 8.6×10^{-5} [eV·K^{-1}].

As an aside, it is of interest to point out that in plasma research or astrophysics one makes a direct connection between the mean kinetic energies of ions and the temperature of a medium.

For example, the sun's surface temperature is about 5800 K or the radiation has an average energy of 0.5 eV.

In plasma devices, one specifies a temperature of about 10^8 K as about 8.6 keV energy.

In stellar conditions, the temperatures are
$T = 10^{10} - 10^{12}$ K $\simeq 0.86 - 86$ MeV.

A particle at room temperature, say T = 300 K, will be of kinetic energy 0.025 eV. It is interesting to note that this kinetic energy is the same for all particles, atoms and molecules, independent of the species. Thus for neutrons of mass $m_n = 939.57$ MeV/c^2, the de Broglie wavelength is

$$\lambda = \frac{1240 \, [\text{eV}][\text{nm}]}{\sqrt{2 \times 0.025 \, [\text{eV}] \times 939.57 \times 10^6 \, [\text{eV}]}} = 0.181 \, [\text{nm}] \tag{1.34}$$

It is of interest to compare the energies of photons, electrons, etc. which will have the same wavelength and thus probe the same dimensions.

For photons:

$$E = \frac{hc}{\lambda} = \frac{1240}{0.181} = 6850 \text{ [eV]} \tag{1.35}$$

$$p = \frac{h}{\lambda} \text{ or } pc = \frac{hc}{\lambda} = 6850 \text{ [eV]} \tag{1.36}$$

For electrons:

$$E = \sqrt{6850^2 + (511 \times 10^6)^2} = 511.046 \times 10^3 \text{ [eV]} \tag{1.37}$$

or photons of 6.85 keV and electrons of 511 keV probe the same dimensions as thermal neutrons of 0.025 eV kinetic energy.

This comparison illustrates that to probe a material of fixed dimensions, say a crystal, momenta should be matched such that their wavelengths are comparable to the sizes under study. Nuclear reactors of high fluxes of thermal neutrons of about 0.025 eV have been workhorses for crystallographic studies in basic and applied sciences. Currently electron synchrotron facilities of several GeV energies provide photon beams of wide spectra of energies and serve the needs of scientific and technological investigations.

1.6 Questions

1. The visible photons are of wavelengths 400–700 nm. Calculate the corresponding photon energies in eV. In X-ray crystallography, photons of energies of about 50 keV are used. Calculate the wavelengths of those photons in [nm].

2. It was asserted that no material object can attain the speed of light. What would be the effect of applying force on an object continuously? Make use of the relativistic mass equation and Lorentz factor (γ) to show that the applied force is not wasted on the body.

3. The masses of electrons (m_e) and protons (m_p) are $m_e = 9 \times 10^{-31}$

kg (0.511 MeV/c^2) and $m_p = 1.673 \times 10^{-27}$ kg (938.3 MeV/c^2), respectively.

These particles are accelerated through potentials of 1 kV, 100 kV and 10 MV.

Calculate the speeds of these particles by Newton's formula and from special relativity.

What do you conclude about the validity of Newton's formula for these cases? To what energy do we need to accelerate protons to attain the speed that an electron acquires when accelerated through 10 MV potential difference?

4. Calculate the de Broglie wavelength of a car traveling at 100 km/h. The car weighs 3000 kg (1 kg $= 0.56 \times 10^{30}$ MeV/c^2).

5. We are designing an experiment to probe the crystal spacing of 0.1 nm. Calculate the kinetic energy of neutrons, electrons, photons and alphas optimized for this experiment. The mass of alpha is $m_\alpha = 3.724$ GeV/c^2.

1.7 Endnotes

Footnote 5: Electron Volt (eV)

In radiation physics and also in atomic, nuclear or particle physics, one employs the electron Volt (eV) as the unit of energy. It is derived from the definitions of Volt [V], Coulomb [C] and Joule [J] as units of potential difference, electric charge and energy, respectively. Accordingly, an object of 1 Coulomb of electric charge will undergo an energy change of 1 Joule as it traverses through a potential difference of 1 Volt. Thus, an entity of one electronic charge $q = e = 1.6 \times 10^{-19}$ Coulomb changes its energy by 1.6×10^{-19} Joules as it traverses a potential difference of 1 Volt.

Footnote 8

Max Planck was the originator of the concept of quantization of energy. In attempts to provide a theory of heat radiation, he had to assume that the energy of a basic element of radiation is proportional to its frequency. The constant of proportionality, which turned out to be a universal constant, is called Planck's constant, $h = 4.135 \times 10^{-15}$ [eV · s] $= 6.626 \times 10^{-34}$ [J · s]. It is not an exaggeration to say that there is no quantum physics without Plank's constant. Physical quantities such as energy, momentum and angular momentum are proportional to this parameter.

Footnote 9

A photon of 1 eV corresponds to an electromagnetic wave of frequency $\nu = \frac{c}{\lambda} = \frac{E}{h} = \frac{1}{4.135 \times 10^{-15}} \frac{eV}{eV \cdot s} = 0.242$ PHz (petaHertz). Or $hc = E\lambda$. As the speed of light is a fundamental constant, $c = 3 \times 10^8$ m/s, the numerical constant of $hc = 1240$ eV·nm is deduced. If the photon energy is changed by a multiple of an electron Volt, its wavelength becomes the corresponding submulitple of a nanometer and vice versa. This fact is indicated in the brackets of Equation 1.16.

Footnote 10

This equation is derived as follows.

A charged particle of electric charge e, velocity $\vec{v}$, moving in an electric field $\vec{E}$ and magnetic field $\vec{B}$ is subject to a force $\vec{F}$, known as the Lorentz force:

$$\vec{F} = e(\vec{E} + \vec{v} \times \vec{B}) \qquad (1.38)$$

If a magnetic field of magnitude B is applied perpendicular to the particle motion and the electric field $E = 0$, the charged particle is subject to centrifugal force. The particle is then bent to follow a circular path of radius of curvature R.

In this arrangement, the magnitude of force is

$$F = evB = \frac{mv^2}{R}$$

or $p = mv = eBR$ and $pc = eBRc$.

From the definition of a magnetic field as Tesla $[T] = [1V \cdot s/m^2]$ we have $pc = eVRB \times 3 \times 10^8 = 0.3BR \times 10^9$ eV.

Here B is given in Teslas and R is in meters. We can simplify this expression as

$$p \left[\frac{\text{GeV}}{c} \right] = 0.3BR \times 10^9 \text{ Tesla} \cdot \text{meters}$$

$$p \left[\frac{\text{GeV}}{c} \right] = 0.3BR \text{ Tesla} \cdot \text{meters}$$

Note that c is suppressed in the equation and shown in units.

2

Radioactivity

2.1 Introduction

Nuclear radiations are ubiquitous. They are present in the atmosphere as cosmic rays originating in outer space, sources of which are yet unknown. They are also present in vegetation as carbon contains a small but finite radioactive isotope,[1] a fact used advantageously to determine the ages of archaeological samples. Most living beings including humans contain calcium, which consists of a tiny amount of long-lived radioactive potassium (^{40}K isotope). The list goes on. In the 20th century, we made significant progress in harnessing energy from nuclear fission, employing nuclear techniques for non-destructive testing of materials, medical diagnostics and therapy. Whether we handle radioactive materials for applications or we are concerned about health and safety, we need to have a good grasp of some basic terminology and be able to do simple calculations to make quantitative estimates of radiation phenomena. When one is concerned about radiation effects, one has to consider the species and energies of the radiations emitted and the activity levels and characteristic lifetimes of the radiation emitting sources. This chapter is devoted to radioactive levels and characteristic times.

2.1.1 Exponential Decay Law

It has been found that in a sample of radioactive material, the intensity of emissions (number of emissions per unit time) decreases exponentially with time. Exponential growths and decays are very common in physical sciences. In the case of growth, we can specify a maximum

---

[1] See Section 2.7.

limit to which a sample, left to itself, will grow after an infinite amount of time. In the case of decay, a sample will vanish only after an infinite amount of time.[2] However, as we will see below, both growth and decay will be quite small after some finite time. For radioactive decays the mean life is a characteristic time specific to the decaying nucleus. To a very good approximation, it does not depend on the physical or chemical state of the material.

Exponential phenomena have an important feature that the decay (growth) rate during any interval depends on the number present at that interval and not on its past history.[3] The decay rate (activity) at any time depends on the number of radioactive nuclei in the sample at that time and not on when, in the past, it was prepared.

We define the "activity" of a radioactive sample as the number of disintegrations per unit time. We commonly use "second" as the unit time. Quite often, minute, hour, day or year is used as the time unit. The activity of a sample depends on the characteristic mean life of the radioactive species and the number of radioactive atoms in the sample.

We also define the "specific activity" of a radioactive material. It is the number of decays per unit time per unit mass. One might use unit mass as 1 gram or 1 kilogram or mol or kmol, depending on the application at hand.

In radiation physics, the terms "mean life" and "half-life" are alternately used. So we define the two.

The exponential decay law asserts that the number of radioactive atoms $N(t)$ and decay rate $A(t)$ at a time t are given by

$$N(t) = N_0 e^{-t/\tau} \tag{2.1}$$

$$A(t) = A_0 e^{-t/\tau} \tag{2.2}$$

where N_0 and A_0 are the number of radioactive atoms and the decay rate at time zero, and τ is a characteristic time.

---

[2]Here we are referring to time as the parameter. Depending on the problem at hand, other parameters such as material thickness may be the relevant one(s).

[3]It must be emphasized that exponential radioactive decay is an experimental law. In physics, the exponential law is given a theoretical support as a random walk problem. A random walk problem describes the chance that a phenomenon depends only on the length of the interval of observation. It is independent of the previous history, i.e., the system has no memory. See the appendix for a brief mathematical account of the exponential law.

Mean life (τ): In a sample of a radioactive material, the number of radioactive atoms and the decay rate would decrease to $1/e \sim 0.368$ or about 36.8% of the original intensity after a lapse of one mean life. It is labeled as the symbol τ (tau).

From the above definition, we see at time $t = \tau$

$$\frac{N(\tau)}{N_0} = \frac{A(\tau)}{A_0} = e^{-1} = 0.368$$

Half-life ($t_{1/2}$): In a sample of a radioactive material, 50% of the radioactive atoms decay and the level of radioactivity decreases to half or a 50% level of intensity after a lapse of one half-life. This parameter is labeled with the symbol $t_{1/2}$ (read as t-half).

The numerical relation between mean life (τ) and half-life ($t_{1/2}$) is $t_{1/2} = 0.693\tau$.

From the half-life definition

$$\frac{N(t_{1/2})}{N_0} = \frac{A(t_{1/2})}{A_0} = e^{\frac{-t_{1/2}}{\tau}} = \frac{1}{2}$$

$$\ln\left[e^{\frac{-t_{1/2}}{\tau}}\right] = \ln\left[-\frac{1}{2}\right] = -0.693$$

$$t_{1/2} = 0.693\tau$$

Decay Disintegration Constant (λ): The inverse of mean life ($1/\tau$) is called the decay constant and it is given the symbol λ. The dimension of λ is inverse time.

From information on the half-life or equivalently the mean life of a radioactive element, we can calculate the number of radioactive atoms in the sample if we know the activity of the sample or vice versa. We can then predict its activity or the number of radioactive atoms in the sample for any future time and trace its past.

Example 2.1

From experiments, it is known that the radioactive level of a ^{60}Co sample decreases to a 50% level in 5 years and a 36.8% ($^1/e$) level in 7.22 years. Thus, the half-life of ^{60}Co is 5 years and its mean life is 7.22 years.

The decay constant of ^{60}Co:

$$\lambda = \frac{1}{7.22}(\mathrm{yr}^{-1}) = \frac{1}{22.727 \times 10^7}(\mathrm{s}^{-1}) = 4.4 \times 10^{-9}\mathrm{s}^{-1}$$

Suppose we have a ^{60}Co atom. The above numbers tell us that the chances that it decays in any one second are 4.4 in a billion. If we keep monitoring this atom, the chances it decays are 0.138 in 1 year, 0.5 in 5 years and 0.632 in 7.22 years, etc. Alternately, in a sample of ^{60}Co, 13.8%, 50% and 63.2% of the atoms would decay, rendering the activity levels 86.2%, 50% and 36.8% in 1, 5 and 7.22 years, respectively.

From the above discussion, we draw the following conclusions:

If a radioactive sample has $N(0)$ atoms at time t_0 and it shows an activity $A(0)$ at that time, the number of atoms $N(t)$ in the sample and its activity $A(t)$ at a later time t are given as

$$N(t) = N(0)e^{-\lambda(t-t_0)} = N(0)e^{-\frac{(t-t_0)}{\tau}} \tag{2.3}$$

$$A(t) = A(0)e^{-\lambda(t-t_0)} = A(0)e^{-\frac{(t-t_0)}{\tau}} \tag{2.4}$$

These equations simply state the fact that after a lapse of one mean life $(t - t_0) = \tau$, the number of radioactive atoms and the activity of the sample both decrease by a factor of $\frac{1}{e} \sim 0.37$.

The ratio of the number of atoms in the sample at any arbitrary time, t, as compared to those at an earlier time t_0, is equal to the ratio of the activities at those times. As time progresses, the number of radioactive atoms decreases and so does the activity of the sample.

Example 2.2

It was reported that the Chernobyl accident of 1986 resulted in the contamination of neighboring countries with ^{137}Cs of a half-life of 30 years. The radiation levels after the accident are listed below (from International Atomic Energy Agency report of the Chernobyl Forum Expert Group 2006)

TABLE 2.1: Areas in Europe contaminated by Chernobyl fallout in 1986. [From Report of Chernobyl Expert Group 'Environment' from Environmental Consequences of the Chernobyl Accident and Their Remediation: Twenty Years of Experience © International Atomic Energy Agency (IAEA) 2006.]

	Area with ^{137}Cs deposition density range (km^2)			
	37–185 kBq/m^2 1–5 μCi/m^2	185–555 kBq/m^2 5–15 μCi/m^2	555–1480 kBq/m^2 15–40 μCi/m^2	> 1480 kBq/m^2 >40 μCi/m^2
Russian Federation	49,800	5700	2100	300
Belarus	29,900	10,200	4200	2200
Ukraine	37,200	3200	900	600
Sweden	12,000	-	-	-
Finland	11,500	-	-	-
Austria	8600	-	-	-
Norway	5200	-	-	-
Bulgaria	4800	-	-	-
Switzerland	1300	-	-	-
Greece	1200	-	-	-
Slovenia	300	-	-	-
Italy	300	-	-	-
Republic of Moldova	300	-	-	-

It will be another 50 and 100 years before the areas with 5–15 μCi/m^2 and 15–40 μCi/m^2 activities will be of radiation levels less than 5 μCi/m^2.

In practical measurements, t_0 is arbitrary. We can set $t_0 = 0$ at the start of a measurement or at the time a sample was prepared. That is what we did in the beginning to define mean life and half-life. With this convention, we may write

$$A(t) = A(0)e^{-t/\tau} \qquad (2.5)$$

Taking the logarithm of both sides, we find

$$\ln[A(t)] = \ln[A(0)] - \frac{t}{\tau} \qquad (2.6)$$

It is interesting to note that the plot of the logarithm of activity versus time yields a straight line slope $-1/\tau$ or decay constant ($\lambda = 1/\tau$).

From the above definitions of half-life and mean life, we recognize

$$\frac{A(t_{1/2})}{A(0)} = e^{-\frac{t_{1/2}}{\tau}} = \frac{1}{2} \text{ or} \qquad (2.7)$$

$$t_{1/2} = \tau \ln[2] = 0.693\tau \qquad (2.8)$$

After, say, n mean lives, the activity decreases to a level $A(n\tau)$ given by

$$\frac{A(n\tau)}{A(0)} = \left(e^{-1}\right)^n = e^{-n} \approx 2.7183^{-n} \approx (0.368)^n \qquad (2.9)$$

Or, after m half-lives, the activity decreases to a level $A(mt_{1/2})$

$$\frac{A(mt_{1/2})}{A(0)} = \left(\frac{1}{2}\right)^m = 0.5^m \qquad (2.10)$$

Figure 2.1 illustrates this feature.

In Figure 2.1, the x-axis shows time in units of mean life (τ). The exponential curve shows the decrease of radioactivity with increasing time. Specifically, the radioactivity is reduced to $0.5^1, 0.5^2, 0.5^3... = 0.5, 0.25, 0.125...$, respectively, of the initial activity at one, two and three half-lives. Also, the radioactivity drops to $0.368^1, 0.368^2, 0.368^3... = 0.368, 0.135, 0.005...$ levels at 1, 2 and 3 mean lives.

From the decay equations (Equation 2.3 and 2.4), it is easy to see that

$$\frac{N(t)}{N(0)} = \frac{A(t)}{A(0)} = e^{-\lambda t} \qquad (2.11)$$

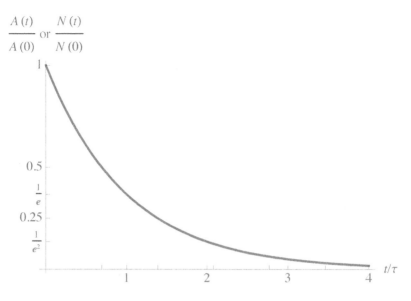

FIGURE 2.1: Radioactive decay: activity and number of atoms versus the time in units of mean lives. Both activity and number of atoms are normalized to 1 at time t_0. The x-axis is plotted in units of mean life (t/τ). Also marked are 1, 2, 3, 4 and 5 half-lives on the same scale. The y-axis shows activity, where $\frac{1}{2}, \frac{1}{e}, \frac{1}{4}, \frac{1}{e^2}$ activity levels are indicated.

Another important relation is the activity[4]

$$A(t) = \lambda N(t) \tag{2.12}$$

The activity of a radioactive sample is equal to the product of the number of radioactive atoms and the decay constant. The longer the mean life, the smaller is the decay constant and thus the sample is less radioactive compared to the same amount of material with a shorter mean life or longer decay constant.

---

[4]This equation is easily derived by calculating the rate of disintegration as

$$A(t) = \frac{dN(t)}{dt} = N(0)\frac{d\left(e^{-\lambda(t-t_0)}\right)}{dt} = -\lambda N(0)e^{-\lambda(t-t_0)} = -\lambda N(t).$$

The negative sign simply indicates that the activity decreases as time increases. In the text above, we suppress the negative sign, as it is of no consequence for our discussion.

To assess the radioactive levels of a sample, we have to ask two questions:

1. How many radioactive atoms are present in a given sample or equivalently what is the level of activity in that sample?

2. What is the decay constant or equivalently what is the mean or half-life of the species?

Example 2.3

In a laboratory experiment, we produced ^{24}Na atoms of 15 hours half-life (mean life $\tau = 21.65$ hours or decay constant $\lambda = 0.0462/h$). We put the source in front of a detector and measured the intensity (counts/min). The table lists the counts/min and ln(counts/min) and the time of measurement of the experiment. The two figures (Figure 2.2) show the plots of measurement time versus counts/min and time versus ln(counts/min). The exponential decay is clearly illustrated. The decay constant as deduced from a fit of counts to exponential function is $\lambda = 0.046$ h^{-1}. The decay constant from a fit of a logarithm of counts to a straight line function is $\lambda = 0.044$ h^{-1}. The two ways of plotting the data are in agreement with the standard value to better than 5% accuracy. Note that this data was taken for about four half-lives in an easy-going way with time intervals chosen at our convenience. It was possible because the lifetime of ^{24}Na was sufficiently but not unduly long. In general, one has to plan the measurement intervals judiciously based on the half-lives and intensities of samples under investigation.

Time [h]	Counts/min	ln (counts/min)
0	28,374	10.253
1.62	23,548	10.067
3.32	22,168	10.006
5.93	21,398	9.971
11.9	12,678	9.448
25.9	7996	8.987
27.5	6594	8.794
48	3084	8.034
60	1962	7.582

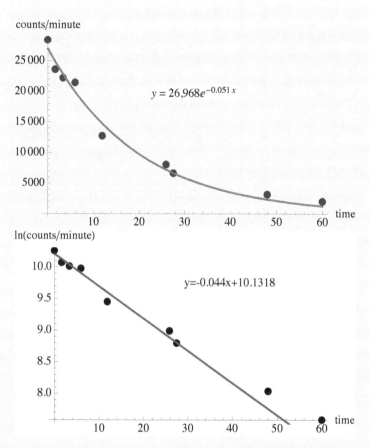

FIGURE 2.2: The decay of ^{24}Na. The table shows the time of measurements and the data. The graphs are counts/min and ln(counts/min) against time from the starting measurement ($t = 0$). The curves are simple fits to data, resulting in disintegration constants 0.046/h and 0.044/h, respectively.

Example 2.4

During the Fukushima reactor incident, radioactivities of ^{137}Cs and ^{131}I and long term effects due to those radiations were of major concern.

^{137}Cs has a half-life of $t_{1/2} = 30.08$ years. Its mean life is $\tau = 43.05$ years. The disintegration constant is $\lambda = 0.023$ yr^{-1}.

^{131}I has a half-life of $t_{1/2} = 8.02$ days. Its mean life is $\tau = 11.57$ days $= 0.0317$ years. The disintegration constant is $\lambda = 0.0864$ d$^{-1} = 31.57$ yr^{-1}.

| After Time | ^{131}I ($t_{1/2} = 8.02$ days) | | ^{137}Cs ($t_{1/2} = 30.08$ years) | |
	Number of Mean lives	Relative activity $\left(\frac{A(t)}{A(0)}\right)$	Number of Mean lives	Relative activity $\left(\frac{A(t)}{A(0)}\right)$
1 Month	2.63	0.072	0.019	0.998
1 Year	31.57	1.9×10^{-14}	0.023	0.977

In one month's time, ^{137}Cs activity is hardly changed, being at a 99.8% level of initial activity, while the ^{131}I activity drops to about a 7% level. After 1 year, ^{137}Cs activity is at about a 98% level of the initial activity while the activity of ^{131}I drops to much less than a trillionth of the initial value during the same time.

2.1.2   Units of Radioactivity

To quantify the measurements, we introduce the units of radioactivity. The international system of units prescribes the unit Becquerel, abbreviated Bq.

Becquerel: A sample is said to be of a radioactivity level of 1 Bq if the rate of disintegration is one decay per second.

Curie: Curie (Ci) is another unit of radioactivity, still in use today.

A sample is said to be of 1 Ci activity if its rate of disintegration is 3.7×10^{10} decays per second.[5]
     Thus, 1 Ci $= 3.7 \times 10^{10}$ Bq.

---

[5]See Section 2.7.

Small plastic sources used in student laboratories are of about 1 microCurie (μCi) or 37 kiloBecquerel (kBq) activities.

Example 2.5

Continuing with the previous example, let us assume that both ^{137}Cs and ^{131}I are of 1 mCi or 37 MBq activities at some initial time, say, t_0.

The number of ^{137}Cs atoms in a 37 MBq activity sample:

$$N(^{137}\text{Cs}) = \frac{\text{Activity in Bq}}{\lambda(\text{s}^{-1})} = \frac{37 \times 10^6}{\frac{0.023}{31,556,926}} = 5.0765 \times 10^{16} \text{atoms}$$

We used the fact there are 31,556,952 seconds in 1 year (365.2425 days).

For the 37 MBq sample of ^{131}I we find

$$N(^{131}\text{I}) = \frac{\text{Activity in Bq}}{\lambda(\text{s}^{-1})} = \frac{37 \times 10^6}{\frac{31.57}{31,556,952}} = 3.698 \times 10^{13} \text{atoms}$$

In one month, ^{137}Cs activity is almost unchanged at 99.8% and the number of ^{137}Cs atoms also is at 99.8% of the initial value.

During the same time, the activity of ^{131}I was reduced to 70 μCi and the number of ^{131}I atoms was reduced to 2.59×10^{12} atoms, about 7% of the initial values.

At the end of 1 year, ^{137}Cs atoms are still abundant at about 97% of initial value, while the ^{131}I has long since disappeared.

It is interesting to note that long lived activities are significant only when the radioactive nuclei are produced in copious amounts. Fewer atoms are needed to produce large amounts of short lived activities.[6] It is comforting to note that short lived activities do not linger for too long. However, one may not conclude that short lives are always better. For many applications, it is necessary to process the materials, transport them to the point of application and carry out the measurements.

---

[6]See Section 2.7.

Thus, one often needs activities with some finite, not too short, half-lives.

It is important to distinguish between the decay rates and radiation levels of nuclear decay. The amount of radiations emitted in a nuclear decay is very specific to the emitting nucleus or the nuclear process under investigation. While decay rates of two sources may be the same, the radiation levels will be higher for that process which emits more radiations per decay. For example, a ^{137}Cs nucleus emits a beta particle and a 661 keV photon per decay. A ^{60}Co nucleus emits a beta particle and two photons of 1173 and 1333 keV energies per decay and thus the radiation levels of ^{60}Co are higher than that of ^{137}Cs for the same decay rates.

^{152}Eu emits several gamma rays per decay and its radiation levels are much higher than those of ^{137}Cs and ^{60}Co of the same decay rates.

2.2   Natural Radioactivity

It is well known that radioactivity was originally discovered in natural samples. Becquerel's alert observation that an exposed photographic film was exposed due to emissions from a uranium sample made history. In a matter of less than 5 years, three types of radiation were discovered. They were called alpha (α), beta (β) and gamma (γ) rays for the first three letters of the Greek alphabet. It turned out that alphas are of positive electric charge and they are nuclei of helium atoms. The betas are negatively charged particles emitted by atomic nuclei. They are, for all practical purposes, identical to atomic electrons. Both alphas and betas are material particles. Gamma rays[7] are electromagnetic radiation, also emitted by atomic nuclei.

It was then recognized that natural radioactivity is a very complex phenomenon. Dedicated works by several researchers identified that there are three chains of decays each with a unique progenitor. They

---

[7]In earlier times, the term gamma rays was specific to electromagnetic radiations from nuclei. We now use this term to refer to photons emitted by nuclei and also rays of very high energies and of cosmic origins.

were identified as 4n, 4n+2 and 4n+3 series.[8] The mass number of each member in the 4n series is divisible by 4, those in the 4n+2 series have a residue 2 and those in the 4n+3 series have a residue 3.

Table 2.2 lists the three series, the head of the chain, its half-life and products.

TABLE 2.2: The three natural radioactive decay series. For each series, the head of the chain, its half-life, end product and number of alphas emitted are listed.

Series	Progenitor	Half-life (billion years)	End product	No. of alphas emitted
4n	Thorium-232	14	Lead-208	6
4n+2	Uranium-238	4.46	Lead-206	8
4n+3	Uranium-235	0.7	Lead-207	7

The figures on the following pages show the decay series of the three natural series along with the artificially produced 4n+1 series, which has ^{237}Np as the progenitor. This progenitor has a half-life of 2.1 million years, while the longest lived activity in this chain is 16 million years (^{233}Pa). These lifetimes are much shorter than the age of earth (~4.5 billion years) and they would be extinct even if they were present in the ancient past.

It is conjectured that ^{239}Pu ($t_{1/2} = 24,100$ y) was the progenitor of the 4n+3 series. Even if it were the case, that memory is lost in the series since the time lapse since the formation of the earth is nearly 190,000 half-lives of ^{239}Pu. This series would have been extinct but for the long life of ^{235}U.

We give a few details of these series below.

---

[8]See Section 2.7.

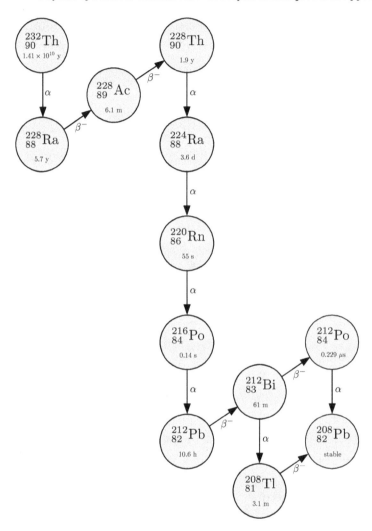

4n series: This is also known as the thorium series. The ^{232}Th with $t_{1/2}$ = 14 billion years is the progenitor of this series. It has gone through 0.32 half-lives. The activity and number of ^{232}Th atoms were reduced to 80% of the initial values at the time of the earth formation. During the series decay, six alpha particles, four beta particles and several gamma rays are emitted by the intermediate decays and the end-product of this series is ^{208}Pb (natural abundance is 52.4%). Here and in the following diagrams, the diagonal arrows (β^- decays) connect isobars of the same mass; the downward lines are α decays.

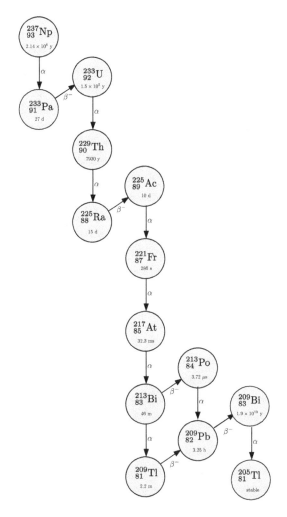

4n+1 series: Conspicuously missing in natural radioactivity is a 4n+1 series. In laboratory experiments, a series with ^{237}Np ($t_{1/2}$ = 2.14 million years) as the progenitor was found. This nucleus is of the longest half-life in this series. This series, if it were present at the time of earth's formation, would have gone through about 2100 half-lives, reducing the activity to less than 10^{-600} of the initial activity. Thus, it is not surprising that such a series would be extinct. The end-product of this series is ^{205}Tl, which is fed by the parent ^{209}Bi with $t_{1/2} = 1.8 \times 10^{19}$ years. As we find ^{209}Bi in natural minerals, this series was likely present in ancient times and has now become extinct.

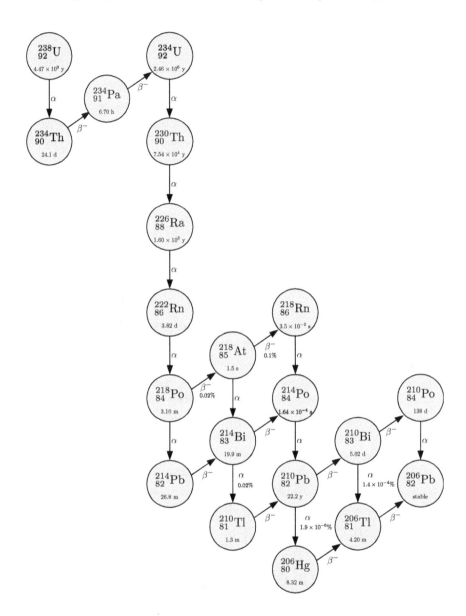

4n+2 series: This is also known as the uranium series. The progenitor is ^{238}U with $t_{1/2} = 4.3$ billion years, just about the age of earth. In this decay chain, eight alpha particles and six betas along with several gamma rays are emitted. This series terminates in ^{206}Pb (natural abundance = 24.1%).

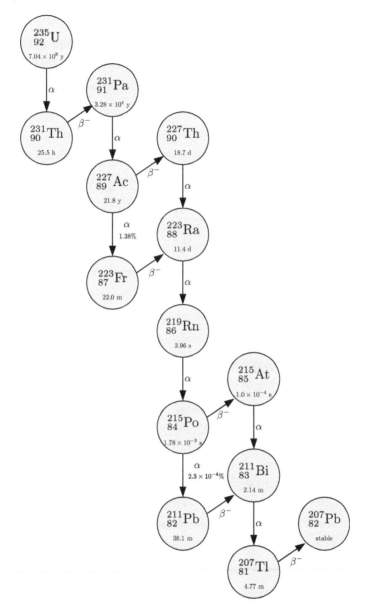

4n+3 series: This series with ^{235}U with $t_{1/2} = 0.71$ billion years as the progenitor terminates in ^{207}Pb (natural abundance = 22.1%). It has gone through nearly 6.4 half-lives, reducing the activity and number of atoms to about 1% of its initial value. In this decay process, seven alpha particles, four beta particles and several gamma rays are emitted.

Today's uranium mineral is composed of 99.3% ^{238}U and 0.7% ^{235}U. If we assume that terrestrial uranium formed at the same time as the earth's formation, 4.5 billion years ago, the relative concentration of ^{238}U atoms and ^{235}U atoms has changed over time due to the different half-lives of the two isotopes.

From the half-lives of ^{235}U and ^{238}U, we have the decay constants

$$\lambda_{238U} = \frac{1}{\tau_{238U}} = \frac{0.693}{t_{\frac{1}{2}}(^{238}U)} = \frac{0.693}{4.47 \times 10^9} \, y^{-1}$$

$$\lambda_{235U} = \frac{1}{\tau_{235U}} = \frac{0.693}{t_{\frac{1}{2}}(^{235}U)} = \frac{0.693}{0.7 \times 10^9} \, y^{-1}$$

The number of atoms in today's sample $N(t = \text{today})$ is the undecayed number.

In passing, we should look at the gamma ray spectrum emitted from a natural uranium sample. Figure 2.3 shows such a spectrum recorded with a high energy resolution HPGe gamma detector. From knowledge based on nuclear physics, we identify the presence of ^{238}U and its descendants.

Some interesting features of the decays of some nuclei are worth noting. For example, look at the decay sequences of the ^{232}Th series. The decay series proceeds through ^{232}Th, ^{228}Ra, ^{228}Ac, ^{228}Th, ^{224}Ra, ^{220}Rn and ^{216}Po by unique paths of alpha and beta emissions to reach ^{212}Bi. The decay first proceeds as

$$^{232}\text{Th} \xrightarrow{\alpha} {}^{228}\text{Ra} \xrightarrow{\beta^-} {}^{228}\text{Ac} \xrightarrow{\beta^-} {}^{228}\text{Th} \xrightarrow{\alpha} {}^{224}\text{Ra} \xrightarrow{\alpha}$$

$$^{220}\text{Rn} \xrightarrow{\alpha} {}^{216}\text{Po} \xrightarrow{\alpha} {}^{212}\text{Pb} \xrightarrow{\beta^-} {}^{212}\text{Bi}$$

^{212}Pb finds two alternate paths of decay to reach ^{208}Pb:

(i) $^{212}\text{Bi} \xrightarrow{\beta^-} {}^{212}\text{Po} \xrightarrow{\alpha} {}^{208}\text{Pb}$

(ii) $^{212}\text{Bi} \xrightarrow{\alpha} {}^{208}\text{Tl} \xrightarrow{\beta^-} {}^{208}\text{Pb}$

The nucleus of ^{212}Bi finds two alternate paths of decay. One path is along ^{212}Bi $\rightarrow$ ^{212}Po $\rightarrow$ ^{208}Pb and the other one is along ^{212}Bi $\rightarrow$ ^{208}Tl $\rightarrow$ ^{208}Pb. The first path is an alpha emission followed by a beta

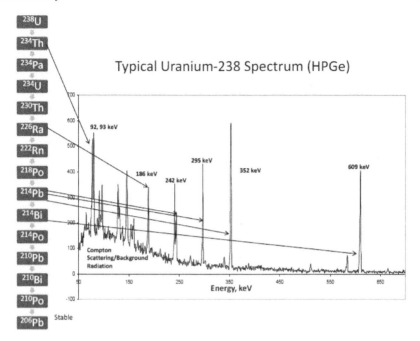

FIGURE 2.3: Typical h hyper pure germanium (HPGe) detector spectrum of a natural uranium sample. Recorded gamma ray energies are used to identify the descendants of the 4n+2 series with ^{238}U as the progenitor.

emission, while the second path is a beta emission followed by an alpha emission. For a single ^{212}Bi atom, we cannot ascertain which path its decay will follow. However, from the literature, we find that a ^{212}Bi atom has 36% and 64% probabilities to decay by emitting alphas and betas, respectively. Thus, a single ^{212}Bi atom has 36% and 64% probabilities to decay along the first and second paths, respectively.

A measurement of a natural uranium ore sample shows gamma rays of energies which can be unambiguously identified with members of the 4n+2 series ^{238}U. In Figure 2.3 we clearly identify ^{234}U, ^{226}Ra, ^{214}Pb and ^{214}Bi isotopes.

These identifications are based on the energies and intensities of gamma ray emissions from the ore sample compared to the information available on the databases, a compilation of analysis of several research works.

In sequential decays, a radioactive atom decays to a product with its characteristic half-life. The unstable decay product then decays to the next in the series with its own characteristic half-life and so on. In radioactivity science vocabulary, we label them parent, daughter, granddaughter and so forth. In this sample, all daughter products are said to be in secular equilibrium with the parent (^{238}U) activity and the intensity changes of emissions follow the half-life of ^{238}U.

2.3   Radioactive Secular Equilibrium

In a chain of sequential decays, relative lifetimes of progenitors and descendants influence the growth and decay rates of descendants. Accordingly, two types of radioactive equilibria are known: transient and secular equilibrium. Below, we discuss the most commonly encountered secular equilibrium in detail.

Secular equilibrium: In a sequential decay, a daughter is in secular equilibrium with its parent if the emissions of the daughter product follow the mean life of the parent. This happens when the parent's mean life (τ_{parent}) is much longer than that of the daughter ($\tau_{daughter}$). For secular equilibrium

$$\tau_{parent} >> \tau_{daughter}$$

or equivalently,

$$\lambda_{parent} << \lambda_{daughter}$$

An analogy might help to understand this.

A water flow rate analogy of radioactive series decays

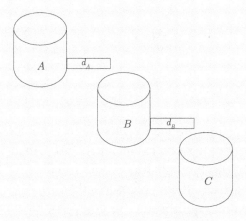

Consider three water barrels A, B and C. At time $t = 0$, both B and C are empty and A is full. We transfer water from A to C through B. A and B are connected by a pipe of diameter (d_A). B and C are connected by a pipe of diameter (d_B). If $d_A >> d_B$, the water flow and rate of filling of C depends on the size of d_B, since B is being filled at a much faster rate than it is being emptied into C. If $d_A << d_B$, then the water flow and rate of filling of C depends on the size of d_A, since barrel B is filled at a much slower rate than the B will drain into C. Simply, the pipe of smaller diameter determines the flow rate. When the rate of draining into C does not depend on d_B, but only on the rate at which barrel B is filled, barrel B is said to be in secular equilibrium with A. Obviously, this situation occurs when $d_B \gg d_A$, i.e., B is readily emptied as it is being filled.

In the same way, for $\lambda_{parent} << \lambda_{daughter}$ the daughter species are populated at a rate much slower than their decay rates and the parent's decay constant determines the decay rate of the daughter. The daughter is in secular equilibrium with the parent.[9]

This feature has many practical uses. In schools and universities, many plastic radioactive sources are used for educational purposes. The most commonly used sources are ^{60}Co and ^{137}Cs. Both ^{60}Co and

---

[9]The mathematics of sequential radioactive decay requires a good first year calculus. The calculation is shown in the appendix.

^{137}Cs are beta emitters, with half-lives of 5 years and 30 years populating levels in ^{60}Ni and ^{137}Ba, respectively. The decays are as follows.

$$^{137}\text{Cs} \xrightarrow[t_{1/2}=30 \text{ years}]{\beta^-} {}^{137*}\text{Ba} \xrightarrow[t_{1/2}=2.55 \text{ minutes}]{\gamma} {}^{137}\text{Ba}$$

$$^{60}\text{Co} \xrightarrow[t_{1/2}=5 \text{ years}]{\beta^-} {}^{60**}\text{Ni} \xrightarrow[t_{1/2}=0.3 \text{ ps}]{\gamma} {}^{60*}\text{Ni} \xrightarrow[t_{1/2}=0.9 \text{ ps}]{\gamma} {}^{60}\text{Ni}$$

where ** and * indicate excited levels in a nucleus, which is unstable. Gamma ray emissions occur much faster than beta emissions. In the case of ^{137*}Ba, the half-life is 2.55 minutes, while those for ^{60}Ni excited levels are less than a picosecond ($<10^{-12}$).

But for secular equilibrium, we could not have inexpensive plastic sources for educational purposes. Also, radiation therapy uses of ^{60}Co would be inconceivable.

Example 2.6

"Moly" is the commonly used radioisotope for medical imaging. A common method of production of this isotope is through the parent ^{99}Mo with a half-life of 65.94 hours. The parent ^{99}Mo emits a beta particle to populate the meta stable level of 142.5 keV energy. The ^{99m}Tc ($t_{1/2} = 6$ hours) is in equilibrium with the parent ^{99}Mo until it is separated by chemical processes.

The metastable ^{99}Tc (labeled ^{99m}Tc) predominately decays by emitting a photon of 140.5 keV, used as a SPECT[10] photon in medical imaging.

On average, a patient is administered 20 mCi of ^{99m}Tc for imaging purposes. As the isotopes are transported across the globe, it takes, say, about 66 hours of transportation, i.e., 11 half-lives of ^{99m}Tc or 1 half-life of ^{99}Mo. If we transport ^{99m}Tc, we must ship $20 \times (2)^{11} = 20 \times 2048 = 41$ Ci of ^{99m}Tc.

To receive the same ^{99m}Tc activity at the destination, we should transport $20 \times 2^1 = 20 \times 2 = 40$ mCi of ^{99}Mo, a ratio of 1000. The longer it takes for transportation to the destination, the larger is this

---

[10]SPECT = Single Photon Emission Computed Tomography.

ratio. Clearly, transporting the daughter in equilibrium with the parent is a more economical, safer, and environmentally friendlier approach. This is the reason that "technetium generators," dubbed "moly cows," are transported around the globe instead of the ^{99m}Tc isotope. In this assessment, we are not considering the losses in chemical separation, etc. and the time it takes to prepare the moly cow after the ^{99}Mo production has stopped.

2.4   Growth and Decay of Radioactivity

There are only a few naturally occurring radioactivities. In laboratories and for applications, one employs artificially produced radioactive substances. A very important criterion is delivery of the optimum amounts of radioactivity to customers economically and effectively. For the production of isotopes, we may bombard a stable isotope with neutrons from a nuclear reactor or particle beams from a particle accelerator. The following reasoning on growth and decay applies to all methods of production. A simple equation contains the entire physics.

We bombard a target material with a beam of particles to result in a constant rate of production R radioactive atoms of disintegration constant λ per second.

During the production time, the number of radioactive atoms grows with time t and they simultaneously decay while they are being produced. The number of radioactive atoms $[N(t)]$ and the activity $[A(t)]$ at a time t are given by

$$A(t) = R \times \left(1 - e^{-\lambda t}\right) \tag{2.13}$$

$$N(t) = \frac{R}{\lambda} \times \left(1 - e^{-\lambda t}\right)$$

$$= R\tau \left(1 - e^{-t/\tau}\right) \tag{2.14}$$

As we continue to bombard the sample, the total number of radioactive atoms and the activity reach saturation values of R/λ and R

at an infinite time. The previous analogy with attempts to fill a leaking water bucket might help to understand this.

If we stop irradiation at a time $t = t_{\text{irradiation}}$ and let the system decay with its exponential decay constant, the number of atoms and decay rates for times $t > t_{\text{irradiation}}$ are given by

$$A(t) = R \times \left(1 - e^{-\lambda t_{\text{irradiation}}}\right) \times e^{-\lambda(t - t_{\text{irradiation}})} \tag{2.15}$$

$$N(t) = \frac{R}{\lambda} \times \left(1 - e^{-\lambda t_{\text{irradiation}}}\right) \times e^{-\lambda(t - t_{\text{irradiation}})} \tag{2.16}$$

$$= R\tau \left[1 - e^{\frac{-t_{\text{irradiation}}}{\tau}}\right] e^{\frac{-(t - t_{\text{irradiation}})}{\tau}} \tag{2.17}$$

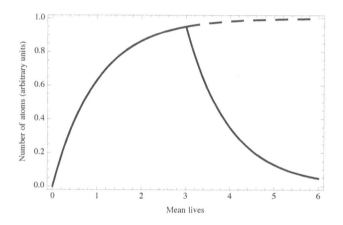

FIGURE 2.4: Plot of ratios of radioactivity and number of atoms to the saturation values plotted against time in units of mean lives.

In Figure 2.4, the horizontal axis is time in units of mean lives (t/τ). The vertical axis represents either

(i) the ratio of numbers of atoms at time t to the maximum number of atoms $\left(\frac{N(t)}{R\tau}\right)$

(ii) the ratio of activity at time t to the saturation activity $\left(\frac{A(t)}{R\tau}\right)$

A single curve represents the growth and decay ratios of the number of atoms and activities. After all, the difference is a factor of the disintegration constant (λ) between the two parameters.

First, for a constant production rate, the number of atoms and activity grow to 63%, 86% and 95% levels of the maximum value, respectively, at one, two and three mean lives. Continuing bombardment for infinite times will result only in the remaining 5% of the production, an enterprise of diminishing returns.

Figure 2.4 shows the growth and decay of radioactivity in this process. In the figure, the solid curve shows the growth reached in three mean lives of irradiation and let decay subsequently. The dashed line shows that the growth will continue to the saturation value of unity on the graph if we continue the irradiation.

Example 2.7

At an isotope production facility, ^{99}Mo is being produced at a rate of 10 mCi/h, while it decays with its characteristic decay constant of

$$\lambda_{99\,Mo} = 0.693/65.94 \text{ h}^{-1} = 0.0105 \text{ h}^{-1}$$

The product ^{99m}Tc decays with a decay constant of

$$\lambda_{99m\,Tc} = 0.693/6.01 \text{ h}^{-1} = 0.1153 \text{ h}^{-1}$$

The growth of ^{99}Mo activity[11] is

$$A_{99\,Mo}(t) = 10 \times \left[1 - e^{0.0105 \times t}\right] \text{ mCi}$$

The growth of ^{99m}Tc activity is

$$A_{99m\,Tc}(t) = A_{99\,Mo}(t) \left[1 - e^{-\left(\lambda_{99m\,Tc} - \lambda_{99\,Mo}\right) \times t}\right]$$
$$= A_{99\,Mo}(t) \left[1 - e^{-0.1048 \times t}\right] \text{ mCi}$$

---

[11]Note that we express the time and disintegration constants in hours and inverse hours, while Curie is unit in seconds. As long as we keep track of multiplication factors carefully, the manipulations will be least cumbersome to get the correct results.

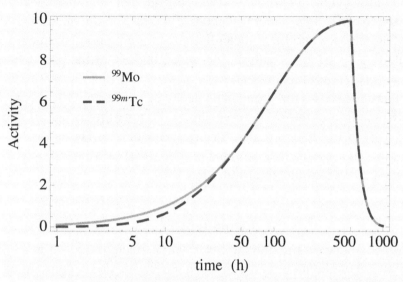

time (h)

The above figure shows the growth of activities of the parent ^{99}Mo and the daughter ^{99m}Tc. For clarity, the time axis is shown in logarithmic scale. A few points are worth noting. Even though we produce the radioactive atoms at a constant rate which corresponds to a 10 mCi decay rate, the growth of activity is very slow and it reaches the maximum value after about five mean lives of the parent. Also, at the early stages of production, the daughter's activity is less than that of the parent. After about three mean lives of the daughter (^{99m}Tc), it is in equilibrium with its parent ^{99}Mo, with the ratio of activities bearing a constant ratio

$$\frac{A(^{99m}\text{Tc})}{A(^{99}\text{Mo})} = \frac{\tau(^{99m}\text{Tc})}{\tau(^{99}\text{Mo}) - \tau(^{99m}\text{Tc})} = \frac{\lambda(^{99m}\text{Tc})}{\lambda(^{99m}\text{Tc}) - \lambda(^{99}\text{Mo})}$$

$$= \frac{t_{1/2}(^{99}\text{Mo})}{t_{1/2}(^{99}\text{Mo}) - t_{1/2}(^{99m}\text{Tc})} = \frac{66}{66 - 6} = 1.1$$

This situation, where the daughter and parent are of comparable lifetimes, with the daughter decaying faster, results in a special case known as "transient equilibrium." In this equilibrium, the daughter is more active than the parent, given by the ratio of the parent's lifetime to the difference of parent-daughter lifetimes, or, equivalently the same ratio of the corresponding half-lives (see Appendix A).

2.5   Age Determination — An Application

The determination of ages of archeological and geological samples as well as those of forensic interest is based on the simple radioactive disintegration equation.

For a sample containing radioactive atoms, left to itself, the rate of disintegration and the number of radioactive atoms present in the sample decrease exponentially in time. If we can measure the specific activity (say, the activity of a sample of unit mass) or the number of atoms of a species of interest, we can determine the age of a sample.

Thus, age determination is based on the following principle. If A_0 and N_0 are the specific activity and the number of unstable atoms of the sample at the time of its formation, then activity $A(t)$ and $N(t)$ after the time lapse of t are given by

$$A(t) = A_0 e^{-t/\tau}$$
$$N(t) = N_0 e^{-t/\tau}$$

We can invert these equations to determine time from the activities or number of atoms as follows:

$$t = \frac{1}{\lambda} \ln \left[\frac{A_0}{A(t)} \right] = \tau \ln \left[\frac{A_0}{A(t)} \right] = \frac{t_{1/2}}{0.693} \ln \left[\frac{A_0}{A(t)} \right]$$

or

$$t = \frac{1}{\lambda} \ln \left[\frac{N_0}{N(t)} \right] = \tau \ln \left[\frac{N_0}{N(t)} \right] = \frac{t_{1/2}}{0.693} \ln \left[\frac{N_0}{N(t)} \right]$$

These two equations guide age determinations by radioisotope measurements. It should be noted that this technique works best when the time scales involved are comparable to the lifetimes of species. If samples are too old, then the activity and species would be extinct or too small to be measurable. If the samples are too new, then the activities and number of atoms in the species are too close to present-day samples. In either case, many precise measurements are warranted.

2.5.1 Radiocarbon Dating

Let us first illustrate the method with well-known radioactive carbon dating. The idea behind this technique is as follows. Live biological specimens, whether living beings or vegetation, constantly exchange carbon with their surroundings, mostly in the form of carbon dioxide. For example, we exhale carbon dioxide while green plants consume carbon dioxide. The samples, once they die, stop this exchange of carbon. From that time on, while the number of stable atoms remains unchanged, radioactive atoms gradually decrease with their characteristic lifetimes. If we can compare the ratio of stable atoms to either radioactivity or the number of unstable atoms of a known disintegration constant, we can estimate, the age of the sample. Natural carbon consists of two stable isotopes, ^{12}C and ^{13}C, the abundances of which are 98.9% and 1.1%, respectively. Present-day carbon also consists of traces of unstable $^{14}C \sim 1.3 \times 10^{-10}\%$, which is constantly produced in the atmosphere as air molecules interact with cosmic radiation consisting of photons, neutrons and charged particles.

For all practical purposes, we can consider the atomic weight of carbon to be due to ^{12}C atoms. Thus, one gram of present-day carbon consists of

$$\frac{\text{Number of } ^{12}C \text{ atoms}}{\text{gram}} = \frac{\text{Avogadro's Number}}{\text{Atomic Weight}}$$

$$= \frac{6.022 \times 10^{23}}{12} = 5 \times 10^{22}$$

From the above, we also have

$$\frac{\text{Number of } ^{14}C \text{ atoms}}{\text{Number of } ^{12}C \text{ atoms}} = \frac{1.3 \times 10^{10}}{98.9} = 1.31 \times 10^{-12}$$

$$\text{Number of } ^{14}C \text{ atoms} = 1.31 \times 10^{-12} \times 5 \times 10^{22}$$

$$= 6.55 \times 10^{10} \text{ atoms/g}$$

The half-life of $^{14}C = 5730$ years

The disintegration constant $\lambda = \frac{1}{\tau} = \frac{0.693}{t_{\frac{1}{2}}}$

$$\lambda = \frac{0.693}{5730} = 1.21 \times 10^{-4} \text{ yr}^{-1}$$

Thus, the specific activity of a present-day carbon sample is

$$A(t) = \lambda N(t) = 1.21 \times 10^{-4} \times 6.55 \times 10^{10} = 1.93 \times 10^6 \text{ yr}^{-1}$$

One rarely carries out measurements for an entire year. One resorts to specific activities per day, per hour, per minute and per second as

$$
\begin{aligned}
1.93 \times 10^6 \text{ yr}^{-1} &= 21,700 \text{ disintegrations/day} \\
&= 904 \text{ disintegrations/hour} \\
&= 15 \text{ disintegrations/minute} \\
&= 0.25 \text{ Bq}
\end{aligned}
$$

We measure the specific activity of a sample of interest and with simple arithmetic we can determine the age of the sample.

Example 2.8

A sample shows ^{14}C specific activity of 10 counts/minute. The age of the sample is

$$t = \frac{5730}{0.693} \ln\left[\frac{15}{10}\right] = 3353 \text{ years}$$

The Shroud of Turin is believed to contain the image of Jesus Christ at the time of crucifixion. If so, the shroud is about 2000 years old. We expect the specific activity of the shroud in the year 2000 A.D. to be

$$
\begin{aligned}
A(t = 2000 \text{ years}) &= 15 \times e^{-2000 \times (1.21 \times 10^{-4})} \text{ disintegrations/minute} \\
&= 15 \times e^{-0.242} \text{ disintegrations/minute} \\
&= 15 \times 0.785 \text{ disintegrations/minute} \\
&= 11.78 \text{ disintegrations/minute} \\
&= 0.2 \text{ Bq}
\end{aligned}
$$

Assuming the sample consists of 1 gram of carbon, we estimate the least precision to which the measurement must be carried out and the corresponding precision in the age determination.

To this end, we proceed as follows: We know

$$t = \tau \ln \left[\frac{A_0}{A(t)} \right]$$

Our interest is to determine the error in age determination due to a measurement error in the specific activity of the sample. The exercise below is a guide to how this is done. The estimates, we should emphasize, are the least precision we should aim for in these measurements. There are several other factors of measurement which will warrant much higher precision.

A little bit of calculus will show that the error dt in age determination of the sample can be written as[12]

$$dt = \frac{\tau}{A(t)} dA \qquad (2.18)$$

and the fractional error in age determination is

$$\frac{dt}{t} = \frac{dA}{A(t)} \times \frac{1}{\ln\left(\frac{A_0}{A(t)}\right)} \qquad (2.19)$$

In the above example, we have

$$\frac{A_0}{A(t)} = \frac{0.25}{0.2} = 1.25,$$

$$\ln\left[\frac{A_0}{A(t)}\right] = \ln(1.25) = 0.223$$

or

$$\frac{dt}{t} = 4.484 \times \frac{dA}{A(t)}$$

Note that the fractional error in the activity measurement must be nearly a factor of 5 smaller than the fractional error we aim for in the age measurement.

In activity measurements, the statistical error is $\Delta N = \sqrt{N}$ and the fractional error is $\delta = \Delta N / N = 1/\sqrt{N}$, where N is the number of measured counts. $N = 1/\delta^2$.

---

[12] See Section 2.7.

% Error in age (age estimate)	Fractional error in activity (δ)	Minimum number of counts required $N = 1/\delta^2$	Estimate of minimum measurement duration per gram of material $= \frac{N}{0.2}$ seconds
10% (200 B.C.E. – 200 C.E.)	0.02	2500	12,500
50% (1000 B.C.E. – 1000 C.E.)	0.1	100	500

The table above shows that a requirement of a factor of 5 (from 50% to 10%) improvement in precision requires that the total accumulated counts be increased by a factor of 25. We may do this by working with more material, longer periods of data taking or a combination of both.

As asserted above, the table shows the least precision required for determining the age to the desired accuracy. There are several other factors which will render the age estimate less precise, first among which is the statistical level of confidence.

The half-life of ^{14}C ($t_{1/2} = 5730$ years) is short in comparison to the geological time scales of a few billion years. As earth scientists engage in the determination of ages of rock formations, they employ various decay chains, each with its range of usefulness.

A search at the NNDC[13] website shows that there are 27 decay chains with half-lives longer than 1 million years. Among them, ^{238}U, ^{235}U, ^{232}Th, ^{129}I, ^{40}K and ^{87}Rb are most commonly used to determine the ages of rocks and minerals. In addition, astrophysicists employ the decay of ^{26}Al as a chronometer of chondrules, knowledge of which is considered essential to understanding our planetary system.

2.5.2 Accelerator Mass Spectrometry

In the 1970s, the direct counting of radioactive atoms was proposed as an alternative to enhance the sensitivities of small samples. As we saw

---

[13]NNDC = National Nuclear Data Center.

in the above example, 10 billion atoms of ^{14}C result in 0.2 Bq. The disintegration rate is the product of the number of radioactive atoms in the sample and the disintegration constant (λ). Thus, samples of longer mean lives (smaller λ) require more radioactive atoms to achieve the same decay rates. If we can count the radioactive atoms, instead of decays, we may be able to get better precision in shorter measurements. This was the idea behind accelerator mass spectrometry.

In the introductory chapter, we noted that charged particles bend in magnetic fields at a radius of curvature given by

$$R[\text{meters}] = \frac{p}{0.3 \times B} \left[\frac{\text{GeV}/c}{\text{Tesla}} \right] \tag{2.20}$$

If we accelerate an ion of the atom in an electric potential V (volts), it gains a kinetic energy T of $T = qV = neV$, where $q = ne$ is the electric charge of the ion. Note that the energy gain depends on the electric charge of the ion and not its mass. Thus, ions of different isotopes in the same charge state will gain the same kinetic energy, but they will have different momenta since[14]

$$p = \sqrt{2mT} \tag{2.21}$$

Thus, in deflecting magnetic fields, the radius of curvature is proportional to $\sqrt{m}$. For ^{14}C and ^{12}C ions in a sample, we will find the ratios of the radii of curvature as

$$\frac{R(^{14}\text{C})}{R(^{12}\text{C})} = \frac{\sqrt{M(^{14}\text{C})}}{\sqrt{M(^{12}\text{C})}} \approx \sqrt{\frac{14}{12}} = 1.08$$

If we arrange detectors to measure ^{14}C and ^{12}C ions simultaneously, we can estimate the age of the sample very directly. We know

$$\frac{\text{Number of } ^{14}\text{C atoms}}{\text{Number of } ^{12}\text{C atoms}}_{\text{present-day}} = 10^{-12}$$

In the sample of interest, the corresponding ratio will be

$$\frac{\text{Number of } ^{14}\text{C atoms}}{\text{Number of } ^{12}\text{C atoms}}_{\text{present-day}} = 10^{-12} \times e^{-\frac{t}{8264}}$$

---

[14]Equation (2.21) is Newtonian mechanics. It is a good approximation for low energies of interest here.

where t is the age of the sample in years.

The measurement can be optimized to maximize the ^{14}C ion counting rate without overloading the system with ^{12}C ions. That is the task of the experimentalist.

The accelerator mass spectrometry can be easily extended to several isotopes besides ^{14}C. Notable among them are ^{36}Cl ($t_{1/2} = 3.01 \times 10^5$ years), of principal interest in hydrology, ^{10}Be ($t_{1/2} = 1.51 \times 10^6$ years) and ^{36}Cl for the investigation of polar ice cores in climatology.

Several other isotopes for different uses in science and technology are being developed.[15]

---

[15] See, for example, a review article by L. K. Fifield in *Electrostatic Accelerators*, p. 461, Ed. R. Hellborg (2005).

2.6 Questions

1. Below we list one of the disintegration parameters (λ, $t_{1/2}$, τ) of a few radioactive elements.

   (a) Calculate the other two parameters (i.e., given λ, calculate $t_{1/2}$, τ etc.). Express lifetimes in multiples of years, days, hours, minutes or seconds so that you have a feeling for how long it survives.

        (i) ^{14}C: $t_{1/2} = 5730$ years

        (ii) ^{22}Na: $\tau = 3.75$ years

        (iii) ^{99}Mo: $\lambda = 2.92 \times 10^{-6}$ s^{-1}

        (iv) ^{56}Co: $t_{1/2} = 77.2$ days

        (v) ^{238}U: $\tau = 6.45 \times 10^9$ years (6.45 billion years)

   (b) We bought a 1 μCi source of each of the above elements. What are the levels of activity after

        (i) 1 week

        (ii) 1 year

        (iii) 10 years

        (iv) 1000 years

2. ^{137}Cs has a half-life of 30 years. We bought a source of one microcurie strength on September 1, 2000. What is the activity of the source on February 1, 2013?

3. Below is a list of some of the commonly used radioisotopes and their half-lives.

Isotope	Half-Life
^{11}C	20 m
^{13}N	10 m
^{18}F	110 m
^{99}Mo	66 h
^{123}I	13 h

As we produce them at a nuclear facility, how long should we irradiate each of the production targets to make 99%, 60% and 20% of saturation activity? Can you comment on the effective use of exposure times?

4. The most common uses of the ^{137}Cs isotope are in undergraduate laboratories, medical applications and other uses due to its emission of 661 keV gamma rays. Its radioactivity can be described as follows. The ^{137}Cs atom emits a beta particle with a half-life of 30 years. As the beta particle is emitted, the nucleus changes to ^{137}Ba and it is populated in an excited level of 661 keV. The excited level is unstable and it emits a photon of 661 keV with a characteristic half-life of 2.55 minutes. The figure shows a sketch of the decay scheme.

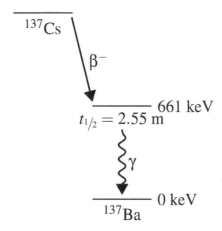

Cesium-137 decay nuclear level diagram

(i) Is the ^{137}Ba in an excited state in equilibrium with its parent ^{137}Cs?

(ii) If we measure the half-life of the 661 keV emitting state with a ^{137}Cs source, what will be the result?

(iii) A chemical separation from a stock sample of ^{137}Cs produces a smaller second sample of ^{137}Ba atoms, which has a measurable activity. What is the measured half-life of the second sample? What will happen to the measured half-life if the separation process is inefficient, resulting in pollution of ^{137}Cs?

5. Our interest in ^{99}Mo is not in the molybdenum isotope but in its daughter product ^{99m}Tc with a half-life of 6 hours.

    (i) After preparation at a nuclear facility, the technetium generator (also known as a moly generator) is transported to a clinic as ^{99}Mo and not as ^{99m}Tc. Give a reason why this is done.

    (ii) Let us say it takes one week for the isotope to reach a clinic from the production point. What are the activities of ^{99}Mo and ^{99m}Tc at the clinic if the activity of ^{99}Mo at the production point is 1 Ci? Assume ^{99m}Tc is in secular equilibrium with its parent ^{99}Mo.

    (iii) Recently there have been proposals to produce ^{99m}Tc (not ^{99}Mo) at nuclear facilities. If you want to have at least 10% of the produced activity reach a clinic, how many hours of handling and transportation of the isotope can you allow for? Compare it with the transportation time of ^{99}Mo.

6. We procure a sample of rock with ^{232}Th and its decay products. By chemical separation, we gradually remove heavier isotopes, starting from ^{232}Th. At each stage of separation, which isotope will determine the time period of radioactivity of this series?

Repeat this exercise for the 4n+2 and 4n+3 series.

2.7 Endnotes

Footnote 1

We identify a chemical element by its atomic number (Z), the number of protons in the atomic nucleus and equivalently the number of atomic electrons in a neutral atom. The atomic number Z of an element determines its position in the periodic table of elements. Atomic nuclei of the same chemical element may be composed of a different number of neutrons and thus be of different weights. In nuclear science, we refer to these distinct atoms of a chemical element as isotopes of the element. This word is derived from the Greek words isos: same, and topos: place or isotope = same place, as they fall in the same place in the Mendeleev periodic table of chemical elements. While the chemical properties of these different atoms are very nearly the same, their nuclear properties are very distinct. More important, just a few of them are stable, and most known isotopes are unstable against decay or emission of radiation. For example, natural carbon consists of three isotopes: ^{12}C, ^{13}C and ^{14}C. Among these isotopes, ^{14}C is radioactive, while the other two are stable. It is a general convention to write an atomic nucleus as $^A_Z X_N$, where Z is the atomic number, the number of protons in the nucleus, which also determines its place in the periodic table of elements. A is the mass number, N is the neutron number and X is the chemical symbol. Quite often N is not indicated as it is trivially given as N = A − Z.

Footnote 5: Curie

Curie (Ci) is a unit of radioactivity, defined as 1 Ci = 3.7×10^{10} disintegrations per second.

If you wonder why we use this awkward number, the unit "Curie" was defined as the activity of 1 gram of radium-226. Radium-226 was the isotope that Madame Marie Curie (1867–1934, a Polish-French scientist) separated by painstaking radiochemical procedures for which she won a Nobel prize.

The mean life of ^{226}Ra is 2309 years. The number of atoms in

1 gram of a substance of mass A is $N = \frac{6.023 \times 10^{23}}{A}$, where we used Avogadro's number.

Thus, 1 gram of ^{226}Ra comprises $N = 2.665 \times 10^{21}$ atoms.

The disintegration constant is $\lambda = 0.000433 \ \mathrm{y}^{-1} = 1.372 \times 10^{-11} \ \mathrm{s}^{-1}$.

The activity of 1 gram of radium is then $A = \lambda \times N = 3.66 \times 10^{10}$ disintegrations/s, which is rounded off to 1 Ci $= 3.7 \times 10^{10}$ disintegrations/s. Therefore 1 Ci $= 3.7 \times 10^{10}$ Bq.

We should note that Antoine Henri Becquerel (1852–1908) was a French physicist and discoverer of radioactivity, for which he shared the 1903 Nobel prize with Madame and Monsieur Curie.

Footnote 6

We consider two samples of two different mean lives $\tau_1 = \frac{1}{\lambda_1}$ and $\tau_2 = \frac{1}{\lambda_2}$, with activities $A_1(t) = \lambda_1 N_1(t) = \frac{N_1(t)}{\tau_1}$ and $A_2(t) = \lambda_2 N_2(t) = \frac{N_2(t)}{\tau_2}$

For equal activities at time t, $A_1(t) = A_2(t)$, the ratio of atoms of species

$$\frac{N_1(t)}{N_2(t)} = \frac{\lambda_1}{\lambda_2} = \frac{\tau_2}{\tau_1} \qquad (2.22)$$

The ratio of the number of atoms is inversely proportional to their mean lives or equivalently half-lives. Thus, a sample with a 1 million year half-life contains one million times as many atoms as that of a sample with a 1 year half-life if both samples have the same activity.

Footnote 8

In decays by alpha emission the residual nucleus (daughter) has the mass number (A) smaller by 4 units, while the mass number of the daughter is the same as that of the parent for both beta and gamma emissions. Gamma emissions are transitions within the same nucleus, while β plus/minus emissions result in a decrease/increase of atomic number (Z), keeping A unchanged. As the natural radioactivities are alpha, beta and gamma, all members of a specific decay chain differ in mass numbers by zero or multiples of four. Thus, the mass number A = 4n+x, where n is an integer and x = 0, 2 and 3 for all nuclei in the 4n, 4n+2 and 4n+3 series, respectively.

Footnote 12

We have $t = \tau \ln\left[\frac{A_0}{A(t)}\right] = \tau \ln[A_0] - \tau \ln[A(t)]$. Among these variables, A_0 and τ are constant since they refer to the mean life of the radioactive species and the specific activity of a present-day sample. Thus, the differentiation is of time (t) and A. From basic calculus, we know that the derivative of a logarithm of a variable x is $d[\ln(x)] = -\frac{dx}{x}$. Thus

$$dt = \tau \frac{dA}{A(t)}$$

or the fractional error (dt/t) in the time determination is given by

$$\frac{dt}{t} = \frac{dA}{A(t)} \times \frac{1}{\ln\left(\frac{A_0}{A(t)}\right)}$$

Q.E.D.

3

Nuclear Energetics

3.1 Introduction

In the previous chapter, we discussed the time variation of intensities of nuclear emissions in radioactive decays. Our concern was only those cases in which decays are possible. Now we must address the question why some nuclei decay while others do not. Also, nuclei decay by emitting a few select types of radiation but not all types. For example, a ^{14}C nucleus with a half-life of 5730 years decays by emitting a beta particle, but it does not emit an alpha particle or a proton or something else. Gamma rays are not emitted. In most cases, daughter nuclei produced in nuclear decays emit gamma rays, but they do not emit neutrons or protons or other complex nuclei. Also, physicists and industry have been producing several nuclei by artificial means with particle accelerators and nuclear reactors. How do they go about it? A further concern is what happens when these emitted particles interact with materials as they travel through a medium. Is it possible that they produce secondary particles such as neutrons, gamma rays or some other particles or some new isotopes? If yes, under what conditions are they produced?

The first and simple answer for all the above questions is the energy balance. Left to itself, a system will tend to the *lowest energy state*, releasing the excess as kinetic energy and radiation. A system, initial or final, may comprise nuclei, particles and/or radiation or some combinations of them. In the total energy balance, we do not distinguish between material particles and photons. However, we make explicit reference to rest masses and kinetic energies of particles. The critical deciding parameter for physical phenomena is the total energy, i.e., the sum of energies due to rest masses and kinetic energies of particles.

Momentum conservation is valid and it is an essential criterion in all considerations.

One important comment is that the energy balance will tell us what is possible, but it will not tell us what happens. However, if the energy balance makes a certain process not possible, then it will not happen. So the energy balance criterion is the first check we do.

3.2   Q-Value of Nuclear Processes

The discussion below applies for the decay of atomic nuclei or that of elementary particles into an ensemble of nuclei, particles and radiation. We may thus use a collective term, bodies, to represent nuclei or particles or radiation without loss of generality. Also, these basic ideas are enough to calculate the energetics of nuclear processes and the production of particles. We try to keep our discussion as general as possible.

For a combination of particles in the initial state and a corresponding set of particles in the final state, we define a Q-value.

The Q-value of a decay or a reaction is the sum of internal energies of bodies in the initial state minus that of the bodies in the final state.

We label E_{initial} and E_{final} as the total energies of the ensemble of particles in the initial and the final state, respectively.

The lowest energy state of a system is one of zero kinetic energies of all the particles.

Consider the decay of a body of mass M_i in the initial state and a final state of the arbitrary number of bodies each of mass M_{1f}, M_{2f}, M_{3f}, etc.

We define[1]

$$Q = \left(M_i - M_{1f} - M_{2f} - M_{3f} - \ldots \right) c^2 \qquad (3.1)$$

Before we embark on detailed examples, one important remark

---

[1] Here we employ the mass-energy equivalence $E = mc^2$ so that the Q-value is given in energy units. For gamma radiation, $E = h\nu$ in energy units is inserted in place of mass.

about nuclear masses should be made. As we deal with small masses and energies, it is advantageous to use multiples of electron Volts as the units of energy. It is also useful to employ atomic mass units (amu's) to express the masses of atomic nuclei. Frequently, Q-values are a very small fraction of the atomic mass unit, being small differences of large numbers. Rounding off numbers will usually give a Q-value of zero, often a wrong answer. Thus, in evaluating Q-values, one has to retain up to the 3rd or 4th significant figures.

3.2.1 Q-Values from Atomic Masses

In laboratories, scientists measure atomic masses and not nuclear masses. Thus, all databases list atomic masses of isotopes.[2]

We may write the mass of a nucleus of atomic number Z and mass number A as:

$$M(A, Z) = M_{atom}(A, Z) - Z \times M_e \tag{3.2}$$

Here $M(A, Z)$ is the mass of the atomic nucleus, $M_{atom}(A, Z)$ is the atomic mass of the same isotope with its Z electrons M_e is the mass of the electron.[3]

For α decays, we can write the equation

$$M(A, Z) \rightarrow M(A - 4, Z - 2) + M(4, 2) \tag{3.3}$$

where $M(A, Z)$ is the mass of the parent nucleus, $M(A - 4, Z - 2)$ is the mass of the daughter nucleus, and $M(4, 2)$ is the mass of the α particle.

As above,

$$M(A, Z) = M_{atom}(A, Z) - Z \times M_e \tag{3.4}$$

$$M(A - 4, Z - 2) = M_{atom}(A - 4, Z - 2) - (Z - 2) \times M_e \tag{3.5}$$

$$M(4, 2) = M_{atom}(4, 2) - 2 \times M_e \tag{3.6}$$

---

[2]See Section 3.8.

[3]Here we neglect the contributions of electron binding energy to the mass of the atom. Some books show that it exists and then drop it from the equations. This is justified since the atomic masses are several GeV, electron mass contributions are a few MeV, while the binding effects are a few keV, about a few parts in a million. Also, as Q-value calculations involve subtraction, it is the difference in binding energies which is relevant. It is an even smaller correction.

or

$$
\begin{aligned}
M(A, Z) =\ & M(A-4, Z-2) + M(4,2) \\
=\ & M_{atom}(A, Z) - Z \times M_e \\
=\ & M_{atom}(A-4, Z-2) \\
& - (Z-2) * M_e + M_{atom}(4,2) - 2 \times M_e
\end{aligned}
\tag{3.7}
$$

The number of electrons on each side are equal, canceling out the M_e terms. So we are left with

$$
M_{atom}(A, Z) \rightarrow M_{atom}(A-4, Z-2) + M_{atom}(4,2) \tag{3.8}
$$

To calculate the Q-value of decays by alpha particle emission, we can work with the masses of neutral atoms found in the data tables. The masses of bare atomic nuclei, devoid of their atomic electrons, do not concern us.

It is thus a simple matter of comparing the atomic masses in the initial and final states.[4] We deduce nuclear masses of interest from the tables as below:

Example 3.1

Consider the alpha decay of

$$
^{226}\text{Ra} \rightarrow\ ^{222}\text{Rn} + {}^4\text{He}
$$

From the tables, the atomic masses (M), in amu, are

Isotope	Mass
^{226}Ra	226.02541 amu
^{222}Rn	222.01758 amu
^{4}He	4.0026 amu

$$
Q = (226.025403 - 222.01757 - 4.0026)\, c^2 = 0.00523 \text{ amu} \times c^2
$$

---

[4]We emphasize this aspect in much detail as we will encounter examples of beta decay by emitting positrons where this does not work.

The first thing to note is the Q-value is positive. Decay by alpha emission is energetically allowed.

1 amu = 931.5 MeV/c^2.

Thus, $Q = 0.00523[\text{amu} \times c^2] \times 931.5[\text{MeV}/c^2] = 4.872$ MeV.

If a nucleus is formed in an excited state, it is heavier than when it is in the ground state.

We can easily verify if a nucleus, say M_{1f}, can be in an excited state of energy E_x (MeV). A quick check is that the Q-value must be higher than the energy of the excited state.

In Example 3.1, levels up to the excitation energies of $E_x = 4.872$ MeV in ^{222}Rn may be populated in the alpha decay of ^{226}Ra. There may be other physics reasons that certain decays will not occur.

3.3 Beta Decay

We consider three types of beta decay:

(i) β^- decay by emission of an electron

(ii) β^+ decay by emission of a positron

(iii) decay by capture of an electron from atomic shells (EC decay)

All these decays are accompanied by emission of a neutrino or antineutrino. An antineutrino accompanies an electron in β^- decay and a neutrino is emitted in decays by electron capture (EC) or position. As a neutrino is of negligible mass,[5] we can safely ignore its presence in the calculation of Q-values for these decays. Decays by positron emission (β^+ decays) and those by capture of an electron result in the same final nucleus. The difference is that during β^+ decay a positron and a neutrino are emitted, while there is no positron in the decay by electron capture. No charged particle is emitted in electron capture.

---

[5]Neutrinos and antineutrinos are the lightest particles. Current estimates of neutron mass are $m_v < 2$ eV, or $m_v < 4 \times 10^{-6} \times$ mass of the electron. They do not carry an electric charge; thus, they are elusive particles, discovered in 1955. While we recognize that they exist, we can ignore them for the purposes of this book.

Below we present the energetics of the three processes.

i. Q-value for β^- decay:

For energy considerations, the process is

$$M(A,Z) \rightarrow M(A,Z+1) + M_e$$

We note that a radioactive atom of zero electric charge with Z protons in the nucleus and Z electrons in atomic shells decays to the final nucleus of Z+1 protons and Z electrons and a free electron. The daughter atom in a positive ion of a (Z+1,A) nucleus has one electron less than a neutral atom.

$$Q = [M(A,Z) - M(A,Z-1) - m_e]c^2 \qquad (3.9)$$
$$M(A,Z) = M_{atom}(A,Z) - Zm_e \qquad (3.10)$$
$$M(A,Z+1) = M_{atom}(A,Z+1) - (Z+1)m_e \qquad (3.11)$$

Therefore

$$Q = ([M_{atom}(A,Z) - Zm_e] - [M_{atom}(A,Z+1) - (Z+1)m_e] - m_e)c^2$$

Thus, Q, in terms of atomic masses, is

$$Q = [M_{atom}(A,Z) - M_{atom}(A,Z+1)]c^2$$

It thus suffices to just check if the mass of the final state atom is higher or lower than the initial state atom. If the initial state atom has higher mass, decay is possible. A simple inspection of the database would do.

Example 3.2

^{60}Co decays by emitting betas (β^-) to ^{60}Ni. What is the highest level of excitation in ^{60}Ni that can be populated in this decay?

^{60}Co has Z = 27 and A = 60; ^{60} Ni has Z = 28 and A = 60. The parent has one less proton and one more neutron than the daughter. The atomic masses are

Initial state:
$$^{60}\text{Co} \ (A = 60, Z = 27) = M_{atom}(A = 60, Z = 27) - 27 \times M_e$$

Final state:
$$^{60}\text{Ni} \ (A = 60, Z = 28) + M_e =$$
$$M_{atom}(A = 60, Z = 28) - 28 \times M_e + M_e$$
$$M_{atom}(A = 60, Z = 28) - 27 \times M_e$$

The difference between the initial and final states is

$$M_{\text{Initial}} - M_{\text{Final}} = M_{atom}(A = 60, Z = 27)$$
$$-M_{atom}(A = 60, Z = 28) \tag{3.12}$$

We need to simply compare the masses of atoms in the initial and final states to see if this decay is energetically possible (i.e., the Q-value is positive or negative).

From the database,

Isotope	Mass
^{60}Co	59.933817 amu
^{60}Ni	59.930786 amu

An atom of ^{60}Co is heavier than that of ^{60}Ni. So the Q-value is positive and decay occurs.

$$Q = (59.933817 - 59.930786) \text{ amu } \times 931.5 \text{ MeV/amu}$$
$$= 0.003031 \times 931.5$$
$$= 2.823 \text{ MeV}$$

Energetically, ^{60}Ni levels up to 2.823 MeV excitation energy may be populated.

ii. Q-value for β^+ decay:

For energy considerations, the process is

$$M(A,Z) \rightarrow M(A,Z-1) + M_{\beta^+}$$

We note that a radioactive atom of zero electric charge with Z protons in the nucleus and Z electrons in atomic shells decays to the final nucleus of $Z-1$ protons and Z electrons and a free positron. A positron and an electron have the same mass (m_e). The daughter atom is a negative ion with one electron more than a neutral atom with a nucleus of $Z-1$ protons. We indicate this by showing the final nucleus as $M^-(A,Z-1)$, a negative ion.

The mass of $M^-_{atom}(A,Z-1) = M_{atom}(A,Z-1) + m_e$,[6] again neglecting the electron binding energy. Therefore, the Q-value is in terms of atomic masses

$$Q = M_{atom}(A,Z) - M_{atom}(A,Z-1) - 2m_e$$

Thus for the decay by positron emission to occur, $Q > 0$,

$$M_{atom}(A,Z) - M_{atom}(A,Z-1) > 2m_e$$

Notice that the difference in the initial and final state atoms must be larger than $2\times$ the mass of the electron for the decay to occur. That is,

$$\Delta M = M_{atom}(A,Z) - M_{atom}(A,Z-1) > 0.0011 \text{ amu} = 1.022 \text{ MeV}/c^2$$

for the positron emission to be allowed.

---

[6]Positrons and electrons are particle-antiparticle pairs, each with the same mass, labeled m_e.

Example 3.3

A positron emitter is ^{22}Na, which decays to ^{22}Ne.

From the data tables,

$M_{atom}(^{22}\text{Na}) = 21.994436$ amu, $M_{atom}(^{22}\text{Ne}) = 21.991385$ amu
Thus,

$\Delta M = 0.003051$ amu $= 2.842$ MeV/$c^2 > 1.022$ MeV/c^2

Therefore positron emission occurs.

In this decay, levels in ^{22}Ne up to excitation energies of $E_x = Q - 1.022 = 1.62$ MeV can be populated. From the nuclear database, we note that only ground states and an excited level at 1.275 MeV are energetically accessible for this decay, as no other excited level in ^{22}Ne exists below 1.62 MeV.

iii. Decay by electron capture (EC):

This is a process in which a parent nucleus (A, Z) captures an electron from its own atomic shells and becomes an (A, Z–1) nucleus. No charged particle is emitted. In this case, our initial state is the atomic nucleus (A, Z) and electrons external to the nucleus.

Thus, the initial state mass is $M_{atom}(\text{A}, \text{Z}) = M_{nucl}(\text{A}, \text{Z}) + \text{Z} \times m_e$.

In the final state, we have a daughter with a nucleus of $(\text{A}, \text{Z} - 1)$ and $(\text{Z} - 1)$ electrons as the system lost an electron, captured by the nucleus.

$$\text{Initial state} = M_{nucl}(\text{A}, \text{Z}) + \text{Z} \times M_e = M_{atom}(\text{A}, \text{Z})$$
$$\text{Final state} = M_{nucl}(\text{A}, \text{Z} - 1) + (\text{Z} - 1) \times m_e$$
$$= [M_{atom}(\text{A}, \text{Z} - 1) - (\text{Z} - 1)M_e] + (\text{Z} - 1)m_e$$
$$= M_{atom}(\text{A}, \text{Z} - 1)$$

Thus

$$Q = M_{atom}(\text{A}, \text{Z}) - M_{atom}(\text{A}, \text{Z} - 1) \qquad (3.13)$$

Similar to the β^- decay process, we can, by a quick inspection of the atomic mass tables, determine whether decay by EC will occur or not. It must be noted that decay by positron emission or EC result in the same final nucleus starting from the same parent state.

It is noted that:

$$Q\text{-value for positron emission} = Q\text{-value for EC} - 2 \times m_e c^2$$
$$= Q\text{-value for EC} - 1.022 \text{ MeV}$$

If the Q-value for EC is greater than 1.022, then decay by either EC or positron emission is energetically possible. If the Q-value for EC is positive and is less than 1.022 MeV, only EC is possible.

If the Q-value is negative, then the initial state is stable against decay by positron emission or EC.

Example 3.4

(i) The decay of zirconium-88 is

$$^{88}\text{Zr} \rightarrow {}^{88}\text{Y} + \beta^-$$

From the mass tables

$$M(^{88}\text{Zr}) = 87.910227$$
$$M(^{88}\text{Y}) = 87.909501$$
$$\Delta M = 0.000716 \text{ amu} = 0.676 \text{ MeV}/c^2$$

The Q = 0.676 MeV is positive but less than 1.022 MeV. Thus positron emission does not occur, but decay of ^{88}Zr to ^{88}Y by EC is possible.

(ii) A second case, the decay of zinc-63

$$^{63}\text{Zn} \rightarrow {}^{63}\text{Cu} + \beta^-$$

$$M(^{63}\text{Zn}) = 62.933211$$

$$M(^{63}\text{Cu}) = 62.929597$$

$$\Delta M = 0.003614 \text{ amu} = 3.366 \text{ MeV}/c^2$$

The Q-value is 3.366 MeV for the transition to the ground state of ^{63}Cu. Also Q is positive for positron emission for levels up to the excitation of $E_{ex} = 3.366 - 1.022 = 2.344$ MeV. Thus levels below 2.344 MeV excitation can be populated by either positron emission or EC, while those between 2.344 MeV and 3.366 MeV can be accessed only by electron capture.

3.3.1 Q-Values from Nuclear Wallet Cards

The NNDC website has a link for nuclear wallet cards.[7] These cards provide a very useful list of isotopes, mass excess of isotopes, their abundances,[8] and also decay modes of unstable isotopes[9] in the format shown below.

Nuclide Z El A	J_π	Δ(MeV)	$T_{1/2}$, Γ, or Abundance	Decay mode

In the first three entries: Z is the atomic number, El is the chemical symbol of the element, and A is the mass number. The cards list all isotopes of a chemical element of fixed Z in increasing mass number A or neutron number (N = A − Z).

The fourth column (J_π) lists the intrinsic angular momentum and parity of the ground state of a nucleus. This does not concern us in energy considerations. The fifth column, Δ, is mass excess. The sixth column is $T_{1/2}$, Γ, or Abundance of a nuclear ground state. For stable isotopes, the entry is the percentage abundance as commonly found in natural samples. For unstable isotopes, the half-life ($t_{1/2}$) or decay

---

[7]http://www.nndc.bnl.gov/wallet/

[8]See Section 3.8.

[9]The decay emissions are denoted by standard abbreviations. Page vi of the wallet card has the key for these notations.

width (Γ) is given. See Remark 3.8 at the end of the chapter for the the correspondence between the half-life and decay width of unstable nuclei. The last column lists the decay modes of the unstable nuclei.

The mass excess (Δ), expressed in energy units, is defined as

$$\Delta = M_{atom}(Z,A) - 931.502 \times A \text{ MeV} \qquad (3.14)$$

The masses are normalized such that the mass of excess of ^{12}C is zero. It comes from the definition of the atomic mass unit.

A few points worth noting are:

- The Q-value calculation is simplified to an easy mathematical addition and subtraction exercise.

- The wallet cards list the mass excess of ground states and metastable[10] states. For excited states, the mass excess is the ground state mass excess plus the excitation energies in MeV.

3.4 Q-Values for Populating Excited Levels

Each atomic nucleus has its own characteristic levels of excitation. The excitation energies at which levels occur and their quantum mechanical properties are very specific to each nucleus. Quite often, radioactivity by emission of an alpha particle or a beta particle results in the daughter nucleus left in an excited level. There would have been no gamma ray emissions if only ground states are populated. The energetics will determine the maximum excitation energy reached in radioactive decay. As always, we must remember that energetics will determine what can be populated but not what will be populated.

We can now define the Q-value for the population of an excited level (Q_{ex}) as

$$Q_{ex} = Q_{g.s.} - E_{ex} \qquad (3.15)$$

It is essential to express all Q_{ex}, $Q_{g.s.}$ and E_{ex} in the same energy units. Here $Q_{g.s.}$ is the Q-value of the ground states, defined above as the Q from the mass differences.

---

[10]See Section 3.8.

As the Q-value for a process to happen must be a positive number, it is easy to see that the maximum excitation energy $E_{ex}(\text{max}) = Q_{g.s.}$. Energy levels of the daughter nucleus above the $E_{ex}(\text{max})$ are not reached by the decay under consideration.

Example 3.5

(i) ^{226}Ra decaying into ^{222}Rn and ^{4}He

For α decay, we find the following data in the wallet cards:
$\Delta(\alpha) = 2.425; \Delta(^{226}\text{Ra}) = 23.669; \Delta(^{222}\text{Rn}) = 16.374$

$$Q = \Delta(^{226}\text{Ra}) - \Delta(^{222}\text{Rn}) - \Delta(\alpha) = 4.87 \text{ MeV}$$

Q is positive and the decay is possible.

(ii) ^{60}Co decaying into ^{60}Ni accompanied by a beta particle and neutrino

From the wallet cards $\Delta(^{60}\text{Ni}) = -64.472$ and $\Delta(^{60}\text{Co}) = -61.649$

Difference of mass excess:

$$\Delta(^{60}\text{Co}) - \Delta(^{60}\text{Ni}) = -61.649 - (-64.472) = 2.823 \text{ MeV}$$

which is the Q-value from Example 3.4.

Note: This is very simple arithmetic. Before we do this, one precaution is to make sure that the sum of mass numbers on the left and right side of the decay or reaction is the same. Also, we check that the sum total of electric charges on the left side is the same as on the right side. That is, the sum of atomic numbers and the charge of electrons or positrons, emitted or absorbed, match on both sides, to guarantee charge conservation.

3.4.1 Nuclear Excitation Energy Determinations

Measurements of energies of radiations has been used as powerful spectroscopic tools to determine true energies of excited states of

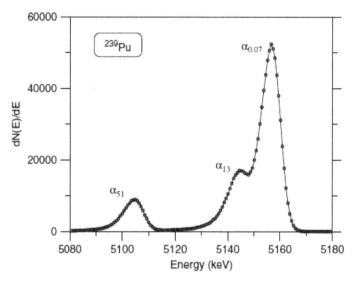

FIGURE 3.1: Energy spectrum of alpha particles emitted in the decay of ^{239}Pu. Three groups of alpha particles, $\alpha_{0.07}$, α_{13} and α_{51}, correspond to excitations of 7 keV, 13 keV and 51 keV, respectively. Taken from Eduardo Garcia-Torano, Applied Radiation and Isotopes 64 (2006) p. 1273.

daughter nuclei. It is straightforward for charged particle decays, as the correspondence is one to one between the charged particle energy and the excitation energy. It is not always that simple for gamma ray spectra, since a cascade of gamma rays may connect two levels via intermediate level excitations.

To illustrate the methodology of deducing the nuclear level information from alpha and gamma spectra, we give two examples. These examples are intended to make it clear that when we see a few distinct energy groups in alpha decays, we can assign a unique level in the daughter nucleus for each alpha group. We can deduce this information from the Q-value of decay and kinetic energies of alpha particles. On the other hand, the gamma spectrum is usually complex and one cannot assign a level for each gamma ray. It takes more involved measurements and some numerical evaluations to arrive at level energy assignments.

Figure 3.1 is an α particle spectrum of the decay of ^{239}Pu to ^{235}U. The Q-value for this decay populating the ground state of ^{235}U is Q

= 5.244 MeV. It clearly shows peaks in the alpha spectrum at three energies, 5157, 5144 and 5105 keV, labeled, respectively, $\alpha_{0.07}$, α_{13} and α_{51}, populating the three excited states at 7, 13 and 51 keV in ^{235}U. Note that this alpha decay does not populate the ground state of ^{235}U, though it is the most favored decay mode.

Example 3.6

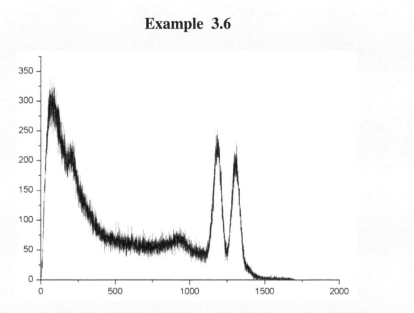

FIGURE 3.2: ^{60}Co gamma energy spectrum. The 1.175 and 1.33 MeV photon peaks are seen right of center. The Q-value of the ^{60}Co decay is 2.823 MeV.

Now we consider a gamma ray spectrum of ^{60}Co decays to an excited state of ^{60}Ni via β^- emission. There is enough energy in the decay to populate levels in ^{60}Ni up to an excitation of 2.823 MeV. We will now look into the database to find the available excited levels of ^{60}Ni, i.e., those levels below the exitation of 2.823 MeV. We find from National Nuclear Data Center data tables that ^{60}Ni has levels at 0, 1.33, 2.158, 2.284, 2.505 and 2.626 MeV. Energetically ^{60}Co can decay to these levels in ^{60}Ni via β^- emission.

From experiments, we observe two gamma rays of 1.175 MeV and 1.33 MeV energies, seen as peaks in Figure 3.2. In ^{60}Ni, there is an excited level at 1.33 MeV but there is none at 1.175 MeV. The sum of the two gamma ray energies is 2.505 MeV, equal to the known excited level at the same energy. We also find that each 1.333 MeV gamma ray is accompanied by one of 1.173 MeV and vice versa. So we easily understand what is happening in these decays. Most of the ^{60}Co decays feed the 2.505 MeV level in ^{60}Ni. It then decays by emitting two gamma rays. The 1.173 MeV gamma ray is emitted as 2.505 MeV level decays to the 1.333 MeV level. In turn, the 1.333 MeV level decays to the ground state by emitting a photon of the same energy. The beta decay of ^{60}Co does not populate any other level. Why this is so has nuclear physics reasons, but it is not of interest to us here.

The figure below is a pictorial representation of ^{60}Co decay for ease of visualization. Only the relevant levels are shown.

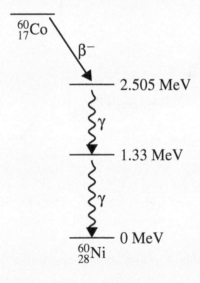

In assigning energy levels from gamma ray spectra, we check energetics to examine what is possible. From there, we take guidance from experiment and describe the observations. In the case of simple decays, such as a single gamma ray emission of ^{137}Cs or two gamma rays for ^{60}Co, the experiments and interpretations are easy. For complicated decays with several gamma ray energies, the experiments are more involved and the interpretations can be difficult. Nearly a half century of studies at various laboratories around the world has produced a wealth of information and it can be easily accessed at the website http://www.nndc.bnl.gov as public domain information.

3.5   Q-Values of Artificial Transmutation

Most experiments of artificial transmutation to produce particles or isotopes consist of hurling projectiles at a stationary target or making two particle beams collide with each other. In these arrangements, there are two bodies in an initial state and an arbitrary number of particles or radiaton in the final state. If two bodies of masses M_a and M_b are present in the initial state and we desire to produce a final state of particles with masses M_{1f}, M_{2f},...etc., we have an equation which is identical to the above relation with $M_i = M_a + M_b$.

The first task is to calculate the Q-value for the process of interest. Q-value calculations amount to the following simple tasks if you are using atomic mass tables:

1. Write down the process for which you want to calculate the Q-value.

2. Make sure that the mass number and electric charge number are conserved.

3. Look up the tables for the masses of particles (a web link is provided).

4. Add up the masses of bodies in the initial state $(M_{initial})$.

5. Add up the masses of bodies in the final state (M_{final}).

6. Calculate the difference $M_{diff} = M_{initial} - M_{final}$.

   - If the result is positive, the process is exoergic and it results in energy release. If the result is negative, the process is endoergic, and energy absorption occurs.

7. Multiply M_{diff} by 931.5 if you are using atomic mass tables.

8. The result is the Q-value in MeV units.

If you are using the table of mass excesses (Δ), the tasks are further simplified.

1. Write down the process for which you want to calculate the Q-value.

2. Check the mass balance and charge balance for the process.

3. Look up the tables for the mass excesses of the nuclei on both sides of the equation.

4. Calculate the difference of the sum of mass excess ($\Delta_{initial} - \Delta final$).

5. The result is the Q-value in MeV units.

3.5.1   Positive Q-Values

In nuclear decays, if the mass of the initial body is higher than the sum of masses of bodies in the final state, Q is positive. Energetically, it is possible that the initial body will change to the final products, with the excess energy liberated as kinetic energy, radiation or other forms of energy.

We are interested in knowing the kinetic energies of products entering detectors or traveling in a medium. In decays with two bodies in the final state, each body is of well defined kinetic energy.

Consider a body of mass M_1, decaying to two bodies of masses M_2 and M_3 with an energy release of Q.

$$M_1 c^2 \rightarrow M_2 c^2 + M_3 c^2 + Q$$

The kinetic energy of the body of mass M_2 (T_2) is[11]

$$T_2 = \frac{M_3}{M_1} Q$$

with

$$T_2 + T_3 = \frac{M_2 + M_3}{M_1} Q \approx Q$$

---

[11] See Section 3.8.

The sum of kinetic energies is slightly less than Q.

To illustrate this feature, let us consider a two body decay involving alpha particles.

Example 3.7

Consider the alpha decay of ^{226}Ra $\rightarrow$ ^{222}Rn + ^{4}He.

From the tables, the atomic masses (M), in amu, are

Isotope	Mass
^{226}Ra	226.025403 amu
^{222}Rn	222.01757 amu
^{4}He	4.0026 amu

$$Q = 226.025403 - 222.01757 - 4.0026 = 0.00523 \text{ amu}$$

Clearly, a rounding off at the first or even second decimal point gives zero as the result and that the decay energy is zero. A wrong result. Indeed, the decay energy is 0.00523 x 931.5 = 4.872 MeV, a substantial release of energy.

The kinetic energy of the alpha particle (T_α) is

$$T_\alpha = \frac{M_{^{222}Rn}}{M_{^{226}Ra}} \times Q = \frac{222.01757}{226.025403} \times 0.00523 = 0.005137 \text{ amu} \quad (3.16)$$

$T_\alpha = 0.0513$ amu $\times 931.502 = 4.786$ MeV

The kinetic energy of ^{222}Rn is

$$Q - T_\alpha = 0.000093 \text{ amu} = 0.086 \text{ MeV} \quad (3.17)$$

The first thing to note here is that the kinetic energies are inversely proportional to their masses. Of the two products, the lighter body carries more kinetic energy than the heavier one. As the two bodies carry the same momentum, their speeds are inversely proportional to the masses.[12]

---

[12]This assertion that the speeds are inversely proportional to the masses is strictly correct only for bodies at non-relativistic speeds. If the speeds of bodies are comparable to that of light, the Lorentz factors (γ) are inversely proportional to their rest masses.

Kinetic energy of the daughter nucleus is very nearly zero. It may be worth pointing here that in alpha decays populating a single level in a daughter nucleus, all alpha particles have a single unique energy. They are mono-energetic. However, in beta decays populating a single level, the beta particles do not have a unique energy. They have a continuum of energies from zero to the maximum value equal to Q-value.[13]

Beta decays are not two body decays, though we considered them to be so for Q-value calculations. We use this information two ways. From known Q-values, we can calculate the kinetic energies of particles and make other measurements; from measured kinetic energies in an experiment, we can deduce nuclear level information. In the decay a neutrino of zero electric charge and of nearly zero mass is emitted.

$$^A_Z a \rightarrow {}^A_{Z+1}b + \beta^- + \bar{\nu}(\text{antineutrino}) \tag{3.18}$$

or

$$^A_Z a \rightarrow {}^A_{Z-1}c + \beta^+ + \nu(\text{neutrino}) \tag{3.19}$$

Now there are three bodies in the final state. Here the mass of a is about the same as that of b or c and much heavier than that of electrons and neutrinos. We can find the sum of the kinetic energies of betas and neutrinos to be

$$T_c = \text{kinetic energy of } \beta + \text{ kinetic energy of } \nu \text{ or } \bar{\nu}$$

$$= Q \cdot \frac{b}{a} \approx Q$$

The fact that a third particle is emitted in the decays affects the kinetic energy of the beta particle. Beta particles do not have specific kinetic energy for a fixed Q-value. Instead, there is a range of kinetic energy values between 0 and Q (T_c).

---

[13]Before the 1930s, physicists observed the continuous spectrum of beta energies. This caused a big problem as it seems not to obey energy and momentum conservation principles. Wolfgang Pauli (1900–1958) came to the rescue by suggesting that there must be a neutrino. It took more than 20 years of research before the neutrino was discovered.

As we illustrated with ^{60}Co, very often a daughter nucleus is left in an excited state, which corresponds to a mass higher than the ground state mass of the nucleus. To calculate the kinetic energies of particles for populating different excited levels, we need to calculate the Q-value for each excited level of energy E_x.

$$Q_{ex} = Q_{g.s.} - E_x \tag{3.20}$$

where $Q_{g.s.}$ is the Q-value calculated from the mass values of the particles involved in the processes.

For each decay channel, we can calculate the kinetic energies of particles. Or, from an experiment we can know the energies of nuclear levels by measuring the kinetic energies of particles. This is especially easy in alpha decays.

For alpha decays, mass of daughter = mass of parent − 4 in amu, or $B = A - 4$. The kinetic energy of alphas feed an excited level $E_x(T_{ex})$.

$$T_{ex} = \frac{A-4}{A} Q_{ex} = \frac{A-4}{A} (Q_{g.s.} - E_x) \tag{3.21}$$

$$or$$

$$E_x = Q_{g.s.} - \frac{A}{A-4} T_{ex} \tag{3.22}$$

The positive Q-value (in units of energy) is the highest excitation that the residual nucleus can be populated in the decays. In the above example, the ^{226}Ra decay can populate levels up to the excitation of $E_x = 4.872$ MeV. Remember that the excited levels are characteristic of a specific nucleus. This realization is a powerful tool for nuclear spectroscopists as they measure gamma ray emissions from radioactive decays and nuclear reactions.

3.5.2 Negative Q-Values

If the mass of the initial body is less than the sum of the final masses, Q is negative. The initial body does not have enough energy to make a transition to the final products.

If we can provide energy from an external agency such as a particle accelerator or nuclear reactor, it may then be possible to make the transformation happen. This is the principle of artificial transmutation. We can create an energetic system by bombarding a target body with energetic photons or other particle beams such as protons, neutrons, alphas, etc. In modern times, particle beams of many nuclear species, unstable particles, photons and electrons of wide ranges of energies are available.

From Equation 3.1 above, we have

$$Q = \left(M_a + M_b - M_{1f} - M_{2f} - M_{3f} - \ldots\right)c^2$$

For a negative Q value, we want to know the threshold energy of the projectile, i.e., the minimum kinetic energy of the projectile to make this process happen.

If a is the target in the laboratory and b is the projectile, the threshold[14] is

$$T_{min}^b = \frac{M_a + M_b}{M_a}|Q| \tag{3.23}$$

If we exchange the roles of the target and the projectile, i.e., if a is the projectile and b is the target, we then have

$$T_{min}^a = \frac{M_a + M_b}{M_b}|Q| \tag{3.24}$$

It is important to note that while the Q-value is the same for a specific reaction, the threshold energy depends on what we use as projectiles in the laboratory (a or b). Lighter projectiles have lower threshold energies for a given combination.

Example 3.8

The PET imaging isotope ^{18}F used as FDG is produced by the reaction

$$^{18}O + p \rightarrow {^{18}F} + n$$

with an accelerated proton incident on an ^{18}O target, resulting in a free neutron and ^{18}F.

The Q-value for this process is

$$Q = \left[M_{initial} - M_{final}\right]c^2$$
$$= -2.438 \text{ MeV}$$

---

[14]The derivation of these relations is given in the appendix.

If we employ proton beams incident on an ^{18}O target, the minimum kinetic energy of protons (T_{min}^p) to cause the reaction ^{18}O$(p,n)^{18}$F is

$$T_{min}^p = |Q| \times \frac{19}{18} = 2.575 \text{ MeV}$$

If, instead, we use an ^{18}O beam onto a hydrogen (proton) target, the minimum kinetic energy of ^{18}O ($T_{min}^{^{18}O}$) to cause $p(^{18}O,n)^{18}$F is

$$T_{min}^{^{18}O} = |Q| \times \frac{19}{1} = 45.997 \text{ MeV}$$

3.6 Separation Energies

Nuclei, when excited to energies beyond a certain minimum, will disintegrate into lighter nuclei accompanied by the emission of a neutron, proton, alpha, etc. The excitation energy of a nucleus at which these emissions become energetically possible is called the separation energy.[15]

For each species, there is a corresponding energy, such as neutron separation energy, proton separation energy, alpha separation energy, etc. for the emission of a neutron, proton, alpha, respectively. Knowledge of these energies is very useful as it could inform us about what energies a specific nucleus will be unstable against the emission of neutrons, protons, etc.

---

[15]Separation energies are the analog to work functions of atoms or molecules. While a single parameter "work function" specifies the ionization of an atom or molecule, nuclear transformations can occur by the emission of a neutron, proton, etc. For each mode, a corresponding separation energy is to be specified.

3.6.1 Neutron Separation Energy

A nucleus (A, Z) becomes a residual nucleus (A − 1, Z) with the emission of a neutron, resulting in the lighter isotope as a product nucleus.

$$(A,Z) \rightarrow (A-1,Z)+n \tag{3.25}$$

For this transformation

$$M(A,Z)+E_x \geq M(A-1,Z)+M_n \tag{3.26}$$

where E_x is the excitation energy and the rest masses, M's, are in energy units. The factor of c^2 is understood.

Interactions of mass excess (E_x):

$$
\begin{aligned}
E_x &\geq M(A-1,Z)+M_n-M(A,Z) \\
&\geq \Delta(A-1,Z)+\Delta n-\Delta(A,Z) \\
&\geq \Delta(A-1,Z)-\Delta(A,Z)+8.071 \text{ MeV}
\end{aligned} \tag{3.27}
$$

on neutron separation energy, S_n, is the minimum value of E_x,

$$S_n = \Delta(A-1,Z)-\Delta(A,Z)+8.071 \text{ MeV} \tag{3.28}$$

3.6.2 Proton Separation Energy

When a proton is emitted, a nucleus (A,Z) changes to another nucleus $(A-1,Z-1)$ or

$$
\begin{aligned}
E_x &\geq M(A-1,Z-1)+M_p-M(A,Z) \\
&\geq \Delta(A-1,Z-1)+\Delta p-\Delta(A,Z) \\
&\geq \Delta(A-1,Z-1)-\Delta(A,Z)+7.289 \text{ MeV}
\end{aligned} \tag{3.29}
$$

The proton separation energy, S_p, is

$$S_p = \Delta(A-1,Z-1)-\Delta(A,Z)+7.289 \text{ MeV} \tag{3.30}$$

3.6.3 Alpha Separation Energy

Another commonly referred to parameter is alpha separation energy, S_α:

$$
\begin{aligned}
S_\alpha &= M(A-4,Z-2)+M_\alpha-M(A,Z) \\
&= \Delta(A-4,Z-2)-\Delta(A,Z)+2.425 \text{ MeV}
\end{aligned} \tag{3.31}
$$

Similarly, we may specify the separation energy for a breakup of a nucleus (A, Z) to fragments $(A - x, Z - y)$ and (x, y) as

$$(A, Z) \rightarrow (A - x, Z - y) + (x, y)$$

as

$$S_{(x,y)} = \Delta(A - x, Z - y) + \Delta(x, y) - \Delta(A, Z) \qquad (3.32)$$

We can determine the separation energies of a nucleus into various combinations of lighter nuclei.

Example 3.9

^{12}C has $\Delta = 0$.
We may consider

(i) The separation energy of ^{12}C to split into three alphas.

From the nuclear wallet card, $\Delta(\alpha) = 2.425$.

The separation energy $(^{12}C \rightarrow 3\alpha)$ is

$$\Delta\left(^{12}C\right) - 3 \times \Delta(\alpha) = -3 \times 2.425$$
$$= -7.275 \text{ MeV}$$

(ii) The separation energy of ^{12}C to split into six deuterons (^{2}H), $\Delta(^{2}H) = 13.136$ is

$$\Delta\left(^{12}C\right) - 6 \times \Delta\left(^{2}H\right) = -6 \times 13.136$$
$$= -157.6 \text{ MeV}$$

3.7 Questions

1. Verify the stability of the ^{238}U nucleus against decay by emission of

    (i) an α particle and ^{234}Th nucleus

    (ii) a β^- and ^{238}Np nucleus

    (iii) a β^+ (positron) and ^{238}Pa nucleus

2. In the decay of ^{231}Pa to an ^{227}Ac nucleus, alpha particles of kinetic energies of 5.058, 5.028, 5.014, 4.951 and 4.736 MeV are observed. Calculate the excitation energies of ^{227}Ac levels populated in these decays.

3. The ^{227}Ac nucleus decays to ^{223}Fr by alpha particle emission. It populates the ground state and excited levels at 0.1, 0.173 and 0.515 MeV. Calculate the kinetic energies of alpha particles feeding each of these levels in ^{223}Fr.

4. To produce ^{99}Tc by the ^{98}Mo(^{3}He,d)^{99}Tc reaction, ^{3}He ions of 20 MeV kinetic energy were used. Calculate the kinetic energies of deuterons (d, heavy hydrogen ion) detected at zero degrees leaving the product ^{99}Tc nucleus in 0.0, 0.18, 0.51 MeV excited levels.

5. Using the nuclear wallet cards data, convert the half-life into width and vice versa for the unstable helium isotopes. Present the results in a table form for the widths and half-lives. Looking at the decay modes, can you draw a general conclusion about the relative half-lives of beta emitters and neutron emitters?

6. Calculate the separation energy of ^{12}C to split into 2 tritons ^{3}H and 2 ^{3}He nuclei.

3.8 Endnotes

Footnote 2

The atomic mass unit (amu) was derived due to a desire to establish a universal reference standard. Carbon was chosen as the reference since it is easily procured and handled. About 99% of naturally occurring carbon is the ^{12}C isotope. The reference ^{12}C mass is defined to be 12 amu, corresponding to mass number 12. The mass of the ^{12}C atom is 11.178 GeV/c^2, yielding 1 amu = 0.9315 GeV/c^2 = 931.5 MeV/c^2. Remember that the energy associated with the ^{12}C atom is 11.178 GeV. There is no need to carry around the c^2 in the calculations.

Footnote 8: Nuclear Wallet Cards

For very short lived isotopes or particles, direct lifetime measurements may not be feasible. Fortunately, another physics principle based on quantum mechanics comes in handy. These levels become broader as the lifetime gets shorter. The widths (Γ) and lifetimes (τ) are related through Heisenberg's uncertainty relation

$$\Delta E \Delta t > \frac{\hbar}{2}$$

This gives

$$\Gamma \tau \approx \frac{h}{2}; \quad \tau \approx \frac{hc}{2\Gamma c} = \frac{1240}{2\Gamma \times 3 \times 10^{23}} \left[\frac{\text{MeV} \cdot \text{F}}{\text{MeV} \cdot \text{F/s}} \right] = 2 \times 10^{-21} \text{s}$$

Half-life $t_{1/2} = 0.693\tau = 1.4 \times 10^{-21}$ s.

For order of magnitude estimates, it is easy to remember as 1 MeV width is $\sim 10^{-21}$ s, 1 keV width is $\sim 10^{-18}$ s and 1 GeV width is $\sim 10^{-24}$ s.

For ^{4}H the table lists a width of $\Gamma = 4.6$ MeV. Accordingly, for ^{4}H

$$\tau = \frac{2 \times 10^{-21} \text{ s}}{4.6} = 4.3 \times 10^{-22} \text{ s}$$

or

$$t_{1/2} = 0.693 \times \tau \approx 3 \times 10^{-22} \text{ s}$$

Footnote 10

In nuclei, meta stable states are excited levels of fairly long half-lives in nuclei. They are also known as isomeric states. These states are likely to decay by emitting gamma rays to the ground state of the nucleus or to the neighboring nuclei by emitting electrons or positrons. In the wallet cards, they are indicated by a suffix m after the mass number. For example, on page 38 of the wallet cards, we find two entries for ^{99}Tc.

Isotopes	Δ	$t_{1/2}$	Decay modes
^{99}Tc	-87.323	2.111×10^{6} y	β^-
^{99m}Tc	-87.180	6.0058 h	IT $\beta^-3.7 \times 10^{-3}\%$

We find the excitation energy of ^{99m}Tc as $\Delta\left(^{99m}\text{Tc}\right) - \Delta\left(^{99}\text{Tc}\right) = 87.180 + 87.323 = 0.143$ MeV.

In decay to the ground state, the radiation of 2.5 keV is accompanied by 140 keV, the most commonly used SPECT photon of ^{99m}Tc with a 6 hour half-life. It is called isomeric transition (IT).

Footnote 11

The kinetic energies of emitted particles are easily derived here. For a body of mass A, decaying to B and C $(A \rightarrow B+C)$, $Q = A - B - C$. For A at rest in the laboratory, the energy balance is

$$A = B + C + T_B + T_C,$$

where T_B and T_C are the kinetic energies of B and C, respectively. As A is at rest, B and C move in opposite directions with equal momenta for the conservation of momentum.

For low energies, we can use Newtonian momentum–kinetic energy relations to write

$$P_B = \sqrt{2T_B B} \text{ and } P_C = \sqrt{2T_C C}$$

We have

$$T_B + T_C = Q \text{ and } T_B = T_C \times {}^C/_B = A\frac{B}{B+C}$$

Thus

$$T_B = Q \times \frac{C}{B+C}$$

In the denominator, we can write $A = B+C$ for cases where $Q \ll A, B$ and C.   Q.E.D.

4

Interaction of Heavy Charged Particles with Matter

In this chapter, we discuss interactions of radiation with matter, which vary with the types and energies of radiation under investigation. The main thrust is on the propagation of radiations, i.e., how far they go in a medium and how much energies they deposit along the way as they pass through a medium. This chapter deals with charged particles heavier than electrons. Electron, photons and neutrons are discussed in the next two chapters.

4.1 Introduction

We cannot overemphasize the need for a good understanding and an intuitive grasp of interactions of radiations with matter. To give some examples, you may be a:

(i) medical physicist assessing radiation effects on living cells as you employ the radiation for diagnostic imaging or therapeutic purposes

(ii) medical physicist estimating radiation interaction effects for medical image reconstruction purposes

(iii) health physicist who addresses radiation doses, or one who needs to determine radiation shielding requirements for occupational safety

(iv) scientist or engineer who designs radiation facilities such as accelerators, reactors or beamlines to deliver radiations

(v) environmental scientist involved in estimating radiation toxicology effects on natural habitats

(vi) space scientist/engineer involved in research with satellites or a space station immersed in harsh cosmic radiation environments

(vii) an electronic or computer engineer developing instruments likely to be employed in high radiation environments

(viii) material scientist determining the radiation hardness of materials for their strength and durability.

There may be many other allied scientists/engineers in industrial or applied science vocations who also need this knowledge.

All we are interested in is how far a charged particle or radiation will travel in a medium and how much energy it deposits along its path. We can provide a detailed answer for an individual heavy charged particle that travels in a medium. In contrast, we can only deduce the attenuation of a beam of neutral radiations or the loss of intensity for them. A loss of intensity is a measure of energy deposited.

For this reason, we divide radiations into

Charged Particles: electrons, muons, pions, protons, alphas, heavy ions, etc. Medical physics and radiation safety scientists encounter these except, perhaps, muons. There was a recent claim that muons also may be used in cancer and AIDS cures.[1]

Neutral Radiations: neutrons, neutrinos and photons (electromagnetic radiations). Neutrinos are relevant only if you are a particle physicist dealing with neutrino physics. So, we do not consider them any further.

There is a fundamental difference in the energy loss mechanisms of electrically charged particles in comparison to that of neutral radiations. Charged particles undergo continuous collisions along their paths and lose energy in a continuous way. Neutral radiations undergo random collisions in a discontinuous manner and their energy loss is not continuous.

---

[1] See US patent http://www.patentstorm.us/patents/6705984/description.html

Thus, for charged particles, we can specify energy loss per distance traveled and range, i.e., the distance they travel in a medium.

Equivalently, for neutral radiations, we can specify loss of intensity per unit distance traveled. It is represented by an attenuation coefficient, a measure of the medium's effectiveness to remove radiations from the main beam of particles.

We must stress that all these parameters (energy loss, range or attenuation coefficient) are specific to the type of radiation and its energy and medium properties.

We further subdivide charged particles into heavy and light particles. Energy loss mechanisms of heavy particles are fairly simple for most practical energies. As they move, they lose energy, undergoing elastic and inelastic collisions with atoms and molecules in the medium and ionizing them. For our purposes, charged particles other than electrons are heavy ones.

For all practical purposes, one single equation, known as the Bethe[2] formula, suffices to describe ionization loss and ranges of heavy charged particles for speeds up to about $v = 0.99c$. Note that it spans a large range of kinetic energies of particles.

Table 4.1 lists the mass energy (also known as rest energy) and kinetic energy[3] of particles of $v = 0.99c$ or $\beta = 0.99$ and $\gamma = 7$.

The Bethe stopping power formula[4] works well up to ultra-relativistic energies, such as for alphas of more than 22 GeV and for gold ions of 1 TeV and beyond.

The Bethe formula is the mathematical expression for the energy loss ($-dE$) of a particle of charge ze moving at a kinetic energy corresponding to a speed of βc as it traverses a small distance (dx) in a material medium characterized by the atomic number (Z).

In Figure 4.1, a particle of charge ze, a mass of m and a speed βc, corresponding to kinetic energy $E = (\gamma - 1)mc^2$, enters a medium of thickness dx, consisting of an element of atomic number Z and mass

---

[2]Hans Bethe (19069–2005). He was awarded a Nobel prize in physics in 1967. His theoretical work spanned nuclear physics, particle physics and astrophysics along with nuclear energy, arms control and science policy. His last publication appeared posthumously in 2007.

[3]See Section 4.8.

[4]See Section 4.8.

TABLE 4.1: Mass energies of far commonly used particles and their kinetic energies for Lorentz factor $\gamma = 7$.

Particle type	Mass energy (mc^2) [MeV]	Kinetic energy for $\gamma = 7$ [MeV]
electron	0.511	3.6
pion	139.5	837
proton	938.3	5630
alpha	3727	22,364
carbon	11,178	182,309
gold	183,474	1,100,843

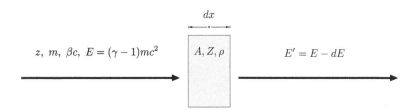

FIGURE 4.1: The energy loss of a charged particle as it traverses a medium. See the text for the meaning of the symbols.

number A. The density of the medium is ρ. On the right, as the particle exits the medium, it has the corresponding speed and momentum for kinetic energy $E - dE$.

4.2   Bethe Stopping Power Formula

The energy loss of a heavy charged particle per centimeter distance it travels in a medium is given by $-dE/dx$ (in units of MeV/cm). The negative sign in the Bethe equation simply indicates that particles lose energy as they travel increasing distances.

$$-\frac{dE}{dx} = 0.024 \times \frac{Z\rho}{A}\frac{z^2}{\beta^2}\left[4\ln\beta\gamma - \beta^2 + 23.02 - 2\ln Z - \delta - U\right] \quad (4.1)$$

Dividing this equation by ρ, the density of the medium, we obtain the mass stopping power $(-dE/dx)$ in units of MeV/(g/cm^2) as

$$-\frac{1}{\rho}\frac{dE}{dx} = 0.024 \times \frac{Z}{A}\frac{z^2}{\beta^2}\left[4\ln\beta\gamma - \beta^2 + 23.02 - 2\ln Z - \delta - U\right] \quad (4.2)$$

In the above equations, z is the ionic charge state of the charged particle ($z = 2$ for alpha particle He^{++}, $z = 1$ for singly ionized He^{+} ions or protons, etc.).[5]

In this form, this equation is applicable to pure elements. We extend the energy loss formula for compounds and mixtures below. The β and γ are the familiar particle speed and Lorentz factor, respectively. The terms δ and U are medium parameters, which can be ignored for general discussion purposes.

In Equation 4.2, the right hand side depends only on the atomic number and mass number. The density, which changes drastically with changes in the physical characteristics of media (gas, liquid or solid) does not appear. Thus, the mass stopping power varies smoothly for materials across the periodic table, and it is independent of temperature, pressure or other extraneous conditions.

ρx has the dimension g/cm^2. Thus, mass stopping power has the dimension energy/(g/cm^2).

Example 4.1

To appreciate the advantage of mass stopping power as a useful parameter, we note that hydrogen gas is of $Z = 1$ and $A = 1$ while a lead medium has $Z = 82$ and $A = 208$, i.e., $Z/A = 1$ and 0.39 for hydrogen and lead, respectively. We then recognize that for fixed ρx, the energy losses are in ratio of 1:0.39 for the lighter hydrogen medium and the heavy lead medium, respectively.

---

[5]See Section 4.8.

Hydrogen can occur in a gaseous state ($\rho = 90\,\mu g/cm^3$) or as a liquid ($\rho = 70\,mg/cm^3$) and solid lead has $\rho = 11.35\,g/cm^3$. A one millimeter thick lead sheet has $\rho x = 1.135\,g/cm^2$. For hydrogen of the same ρx, we need $1.135/7 \times 10^{-2}$ cm = 16.2 m of liquid hydrogen or a $1.135/9 \times 10^{-5}$ = 12.6 km hydrogen gas column.

It is important to note that this formula is parameterized by the speed of the particle in a medium. That is, the pertinent parameter is how fast a particle is moving in the medium and how much it "sees" of particles in the medium, which determines the energy loss of a particle in a medium.

A close inspection of the above equations reveals some interesting features.

1. The term $\rho Z/A$ describes the medium in which particles are moving. The elemental aspects are contained in Z/A, the ratio of atomic number and mass number of the medium. Note also there is a $\log Z$ term which is due to ionization energies of the medium. A further small contribution is in δ and U, which are model dependent corrections and we ignore them, especially when our interest is not in details but a general estimate of stopping power. Information about the physical state (gas, liquid or solid) of the medium is contained in the material density (ρ). As discussed above, mass stopping power does not depend on the density of the medium.

2. The properties of a moving particle are contained in z, the ionic state of moving charged particle, and its speed, β. As a particle loses energy, its speed decreases and thus the energy loss rate changes. Also, a particle may lose or gain charges as it interacts in the medium. Thus, z is not always the atomic number of the moving particle. It is the charge state of a particle in the medium. One resorts to statistical methods to estimate those probabilities and estimate energy losses. We often carry out numerical integrations with the aid of computer programs to determine the energy loss over a finite distance or the total distance that a particle travels in a medium.

3. We also note that the equation says that all material particles of the

same charge state moving at the same speed lose the same amount of energy in a medium.

For particles with the same speed, the energy loss rate varies as the square of the ionization state of the particle (z^2).

Helium ions can occur as alpha particles of doubly charged ($z = 2$) or singly charged ($z = 1$) ions. For the same kinetic energy, alpha particles lose four times the energy as singly ionized helium.

In Table 4.1 we list several particles and their kinetic energies, all of same $\beta = 0.99$. The alpha particle is a doubly charged particle of $z = 2$, while pions and protons are of $z = 1$. We can make singly ionized carbon, gold, etc. ions of $z = 1$. Carbon, gold, etc. ions tend to lose or gain electrons as they move in a medium and thus change their charge states. One has to include those effects to calculate energy losses.

We then conclude that protons, pions, singly ionized carbon and gold all of same β but of different kinetic energies undergo the same energy loss rate as they pass through a small region of a material medium. But when they lose the same amount of energy, the speeds change for different masses of particles. Thus, as different charged particles of initially the same speed traverse the same amount of a material medium, they emerge with different speeds at the end. Their energy loss rate changes as they propagate in the medium and thus the overall distance they travel in a medium also is different for different masses.

Stated differently, when we deal with particles of the same kinetic energy but different masses, the stopping power[6] of heavier ones with smaller β is higher than that of lighter ions of larger β.

Figure 4.2 is a plot of the variation of stopping power ($-dE/dx$) for $z = 1$ plotted against $\beta\gamma$ (the product of speed/c and the Lorentz factor) in arbitrary units. For clarity, we choose to plot $\beta\gamma$, which is approximately equal to γ for large β, and it increases to infinity as β approaches unity.

Note that both the $\beta\gamma$ and the $-dE/dx$ axes are logarithmic scales.

---

[6]Stopping power is the rate at which a particle loses its energy as it traverses through a medium. Media of higher stopping powers are more effective in containing the effects of radiation in smaller volumes. For experimental arrangements, media of higher stopping powers are not necessarily better. The choice of media and particles depends on the applications.

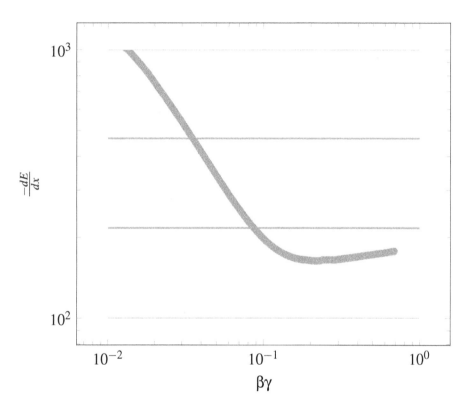

FIGURE 4.2: Stopping power of a singly charged particle ($z = 1$) versus $\beta\gamma$ of the particle.

We may consider this to be a universal curve of energy loss versus momentum for all charged particles.

4.2.1 Minimum Stopping Power

We note that stopping power is very high for low β and thus at low energies. For increasing speeds, the stopping power decreases, reaching a shallow minimum at about $\beta = 0.95$ ($\beta\gamma \sim 3$) and becomes nearly constant for higher energies.

From this curve, we can calculate the energy loss for all particles of arbitrary speeds and corresponding momenta, since the momentum of a particle is $p = m\beta\gamma$ in energy/c units. Here the mass m is expressed in energy/c^2 units, usually MeV/c^2 or GeV/c^2 units.

For example, at minimum stopping power, particles lose the least amount of energy. Minimum stopping power corresponds to the energy or momenta where the charged particles cause minimum ionization. Momenta and the corresponding kinetic energies of some particles at the minimum ionization condition are given in Table 4.2.

TABLE 4.2: Momentum and kinetic energy of a few minimum ionizing particles ($\beta\gamma = 3$).

Particle	Momentum [MeV/c]	Kinetic Energy [MeV]
electron	1.533	1.022
muon	315	210
pion	418.5	279
proton	2815	1876
alpha	11,260	7506

From Table 4.2 we see that the heavier the particle the higher is the momentum/energy at which it is minimum ionizing. For example, protons are minimum ionizing at about 1.9 GeV kinetic energy, while an electron is minimum ionizing at about 1 MeV. The minimum ionizing energy scales linearly with the rest mass of the particle. Simply, express the mass of a particle in energy units and multiply it by two. This gives us the minimum ionizing energy of a particle. The kinetic energy for which a particle is minimum ionizing is twice the rest energy of that particle.

For example, a kaon is a charged particle of mass energy of 493.5 MeV. It becomes minimum ionizing at about 987 MeV or about 1 GeV kinetic energy. The total energy of a kaon is about 1.5 GeV.

Example 4.2

Calculate the kinetic energy at which carbon atoms are minimum ionizing.

Charged particles are minimum ionizing when $\beta\gamma = 3$ or

$$\beta \cdot \frac{1}{\sqrt{1-\beta^2}} = 3$$

yielding $\beta = 0.949$ or

$$\gamma = \frac{1}{\sqrt{1-\beta^2}} = \frac{3}{0.949} = 3.162$$

The kinetic energy is

$$E_k = (\gamma - 1)mc^2$$

For ^{12}C ions, rest energy (mass) $= 12 \times 931.5 \text{ MeV}/c^2$

$$= 11{,}178 \text{ MeV}/c^2.$$

or

$$E_k = (\gamma - 1)mc^2 = 2.162 \times 11{,}178 \text{ MeV}$$
$$= 24{,}170 \text{ MeV}$$
$$= 24.17 \text{ GeV}$$

The total energy E of ^{12}C ions = kinetic energy + rest energy
$$= 35.35 \text{ GeV}$$

The momentum of these ^{12}C ions $= \dfrac{E\beta}{c} \dfrac{\text{GeV}}{c}$
$$= 35.35 \times 0.95 \frac{\text{GeV}}{c}$$
$$= 33.58 \frac{\text{GeV}}{c}$$

We note that electrons become minimum ionizing very quickly and behave quite distinctly from other charged particles. We discuss these details in the next chapter.

4.2.2 Relative Stopping Power

Let us explore this equation a bit further. We note, for a given medium

$$-\frac{dE}{dx} \propto \frac{z^2}{\beta^2} \qquad (4.3)$$

At low speeds, we simply use Newton's expression for kinetic energy being proportional to the square of a particle's speed.[7]
Or

$$-\frac{dE}{dx} \propto \frac{mz^2}{K.E.} \qquad (4.4)$$

where m is the mass of particle.

If we now compare the energy loss rates of two particles of charges z_1, z_2, masses m_1 and m_2, and of kinetic energies T_1 and T_2, we find

$$\frac{\left[\frac{dE}{dx}\right]_1}{\left[\frac{dE}{dx}\right]_2} = \frac{\frac{m_1 z_1^2}{T_1}}{\frac{m_2 z_2^2}{T_2}} \qquad (4.5)$$

For $T_1 = T_2$

$$\left[\frac{\frac{dE}{dx}_1}{\frac{dE}{dx}_2}\right]_{T=fixed} = \frac{m_1 z_1^2}{m_2 z_2^2} \qquad (4.6)$$

For example, alpha particles are approximately four times as massive as protons and they carry twice the amount of charge ($m_1 = 4m_2$ and $z_1 = 2z_2$).

We infer that for the same kinetic energy, the energy loss of alphas ($z_1 = 2, m_1 = 4$) is 16 times as much as that of protons ($z_2 = 1, m_2 = 1$).

We may estimate the kinetic energy of alphas at which they lose the same energy as protons. For kinetic energies of α particles (T_1) and that of a proton (T_2)

$$[dE/dx]_\alpha = [dE/dx]_p \qquad (4.7)$$

---

[7]Note that in this low energy approximation, the speed squared and kinetic energy are proportional to each other. But this strict proportionality breaks down at relativistic speeds.

It is given by

$$\frac{m_1 z_1^2 T_2}{m_2 z_2^2 T_1} = 1 \text{ or } T_2 = \frac{m_2 z_2^2}{m_1 z_1^2} T_1$$

i.e., for the same energy loss as protons, alphas will have 16 times the kinetic energy of protons. Lower energy alphas undergo higher energy losses and they travel shorter paths. Their range is shorter.

Quite often the energy losses and ranges of ions are compared to those of protons with $m = 1$ and $z = 1$. To this end, we can rewrite the above equation as

$$\frac{dE}{dx_{ion}} = M_{ion} z_{ion}^2 \frac{T_p}{T_{ion}} \frac{dE}{dx_p} \tag{4.8}$$

or for $T_p = T_{ion}/M_{ion}$,

$$\frac{dE}{dx_{ion}} = z_{ion}^2 \left[\frac{dE}{dx} \, p \right]_{T_p = \frac{T_{ion}}{M_{ion}}} \tag{4.9}$$

For example, an alpha particle ($z = 2$) will have 4 times the energy loss as a proton of $1/4$ the kinetic energy of the alpha particle.

Example 4.3

Calculate the stopping powers of singly ($^{12}C^+$), doubly ($^{12}C^{++}$), and triply ($^{12}C^{+++}$) charged carbon ions of 1 GeV kinetic energy.

Hint: first find the stopping power of protons of corresponding energy.

$$\left. \frac{dE}{dx}(Z, A) \right|_T = z^2 \times \left. \frac{dE}{dx}(proton) \right|_{\frac{1}{A}T}$$

Approximating the mass of a proton as 1 amu, we have to find the energy loss (stopping power) of the proton with a kinetic energy of 83.333 MeV:

$$T_p = \frac{T_C}{M_C} M_p; \quad \frac{T_C}{12} \text{ [MeV]} = \frac{1000}{12} \text{ [MeV]} \sim 83.333 \text{ [MeV]}$$

From PSTAR, the mass stopping power in water of a 1 GeV carbon ion with a +1 charge is

$$\frac{dE}{dx}(1,12)\bigg|_T = 1^2 \times \frac{dE}{dx}(proton)\bigg|_{\frac{1000}{12}MeV} = 8.387 \left[\frac{MeV}{cm}\right]$$

The linear stopping power in water for the doubly ionized 1 GeV carbon ion is

$$\frac{dE}{dx}(1,12)\bigg|_T = 2^2 \times \frac{dE}{dx}(proton)\bigg|_{\frac{1000}{12}MeV} = 33.548 \left[\frac{MeV}{cm}\right]$$

The linear stopping power in water for the triply ionized 1 GeV carbon ion is

$$\frac{dE}{dx}(1,12)\bigg|_T = 3^2 \times \frac{dE}{dx}(proton)\bigg|_{\frac{1000}{12}MeV} = 75.483 \left[\frac{MeV}{cm}\right]$$

In atomic beam experiments, one produces ions of different charge states. For example, ^{4}He can be ionized to become an alpha particle ($z = 2$) or a singly ionized ^{4}He ion ($z = 1$). We note that alpha particles lose four times the energy of a singly ionized ^{4}He, when they are of the same kinetic energies. Similar conclusions can be drawn for ions of all species.

One also encounters ions of the same mass but of different chemical elements. For example, doubly ionized ^{3}He loses four times as much energy as an ^{3}H ion of a single charge. However, the energy loss of singly ionized ^{3}He and that of an ^{3}H ion are identical.

These observations are very useful in the transport of charged particles through material media and the design of experiments concerned with identification of charged particles.

In several experimental arrangements at particle accelerators, one produces charged particles of different species, say proton, deuteron, pion, muon, etc. By sending them through magnetic fields, one selects all charged particles of the same momenta but of different kinetic energies. If one measures the energy deposits in particle detectors, providing the information of energy loss, $-dE/dx$, in detector thickness, dx, one has enough information to determine the masses of individual particles and identify them. This technique works well at low energies, below the minimum ionizing region.

Because for fixed momentum, each species of particles has a different energy loss, knowing the momentum and the energy loss, we can identify the particle. Above minimum ionization energies, energy losses are not significantly different; this method of identification is more difficult, though it has been employed with some good results.

To close this section, it may be worthwhile to look at another figure (Figure 4.3), where we plot the energy losses of electrons (e), muons (μ), pions (π), kaons (K) and protons (p) of masses 0.511, 105.0, 139.5, 493.5 and 938.3 MeV/c^2, respectively, against momentum ($p = m\beta\gamma$). In this figure, the solid curves are theoretical lines. Superimposed on these curves are the experimental data of energy loss measurements, shown as scattered dots. Two conspicuous features are noteworthy. For low momenta, the energy loss falls on a distinct curve for each particle species, displaced from one another by the mass factor. The lighter particles become minimizing at lower momenta. The figure shows that minimum ionization occurs for pions and muons at above 0.3 GeV/c, for kaons at 0.9 GeV/c and for protons at about 1.3 GeV/c. Above the minimum ionizing region, energy losses are nearly independent of the particle mass. In the plot, we see that above about 2 GeV/c momenta, the experimental data shows a big clutter of points touching all theoretical curves. It also means that energy loss measurements are not the best tools to identify particles at very high energies. In recent years, groups at CERN have shown that careful measurements at very high energies might help in particle identification, since the energy loss plateau for each particle has a different magnitude. Over the past decades, physicists have developed various physical methods to identify particles over wide energy ranges.

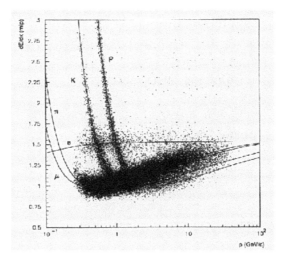

FIGURE 4.3: Energy losses of electrons (e, $m_e = 0.511$ MeV), muons (μ, $m_\mu = 105$ MeV), pions (π, $m_\pi = 139.5$ MeV), kaons (K, $m_K = 493.7$ MeV) and protons (p, $m_p = 938.3$ MeV/c) versus momentum. The lines are Bethe's formulas and the points are experimental data. [From J. Beringer et. al. (Particle Data Group), Phys. Rev. D86, 010001 (2012).]

4.3 Range of Charged Particles

For heavy charged particles and low energy electrons, one can estimate the distance a particle of given kinetic energy will travel in a medium, known as the range of the particle in the medium. This depends on the particle species, its kinetic energy and the medium's properties.

As we have seen above, a medium is specified by ρ, Z and A factors in the equation. The range of a particle of a specific kinetic energy corresponds to the distance it travels for the corresponding β (γ). At the end of the range, the particle comes to rest, or $\beta = 0$ and $\gamma = 1$, and the kinetic energy is zero.

Mathematically, the range R of a particle is given by the integral

$$R = \int_{E_0}^{0} \frac{dE}{\left(dE/dx\right)} \qquad (4.10)$$

where dx is an infinitesimal distance the particle of energy E travels while losing energy dE. A particle starting at kinetic energy E_0 has final energy zero as it comes to rest.

We make the following observations:

(i) For a given kinetic energy, heavier particles move slower, (dE/dx) is larger for smaller β or larger m, thus

$$R \propto \frac{1}{m}$$

(ii) We have seen $(dE/dx) \propto z^2$, z is the ion charge, or

$$R \propto \frac{1}{z^2}$$

Combining the two, we can write

$$R \propto \frac{1}{m_{ion} z_{ion}^2} \qquad (4.11)$$

From this proportionality, we can estimate the ranges of all charged particles of non-relativistic energies from that of a reference particle.

For reasons of convenience, the ranges of protons are used for reference. Range R_{ion} of mass m_{ion} and charge z_{ion} is compared to the range R_p of the same kinetic energy.

$$\frac{R_{ion}}{R_p} = \frac{m_p z_p^2}{m_{ion} z_{ion}^2}$$

$$= \frac{1}{A z_{ion}^2}$$

where A is the mass number of the ion.

For example, an α ($A = 4$, and $Z = 2$) has a range

$$R_\alpha = \frac{1}{16} R_p$$

For an α and a proton of the same kinetic energy.

It is also easy to see that ions of the same mass and of different charge states travel ranges inversely proportional to the square of their charges.

For example, doubly ionized ^{3}He and singly ionized tritium (^{3}H), both of same mass, travel

$$\frac{R_{^3\text{He}}}{R_{^3\text{H}}} = \frac{z^2_{^3\text{H}}}{z^2_{^3\text{He}}} = \frac{1}{4}$$

^{3}He ions travel one quarter the distance of tritium ions of the same kinetic energy. Further, as with stopping powers, we can approximate

$$R_{ion} \propto \frac{T_{ion}}{m_{ion}z^2_{ion}} \qquad (4.12)$$

4.4 Bragg Peak

A characteristic feature of charged particle propagation is maximum energy loss in a short distance, occurring at the end of the particle trail, known as the Bragg[8] peak. This feature is heavily exploited in radiation therapy. The occurrence of the Bragg peak is intuitively obvious. A look at the energy loss distribution (Figure 4.2) shows that, for energies below minimum ionization, energy loss increases as the energy of a particle decreases. So, if we start with a particle of some fixed kinetic energy, the energy loss per unit length increases as the particle slows down along its path in the deeper medium. As an ion loses more and more energy, the energy loss decreases and becomes zero at the end. A particle, after all, cannot lose more than its own kinetic energy. Thus, a particle, as it traverses a medium, experiences increasing energy loss in the interior, followed by decreasing energy before it comes to rest. This phenomenon manifests as the Bragg Peak.

Toward the end of its journey, near the range as it comes to a stop, a

---

[8]Sir William Henry Bragg (1862–1942) Australian/British physicist. His studies around 1904 led to the discovery of the Bragg peak. He shared the Nobel Prize in Physics with his son William Lawrence Bragg (1890–1971) for their services in the analysis of crystal structure by means of X-rays. Almost everyone would be familiar with Bragg's law of diffraction: $n\lambda = 2d \sin \theta$, where λ is the wavelength of radiation, d is the aperture width, and θ is the angle of nth order diffraction.

particle will have very little energy to lose. In a few encounters it may have less energy than the thermal energies of its collision partners and it may gain a bit of energy.

This will cause particle straggling, i.e., not every particle of the same energy stops at the same depth in the medium, and there is a small but finite depth over which the particles stop.

The range and depth at which the Bragg peak occurs depend on the particle type, its energy, and the medium in which it travels. Radiation therapy of tumors with ions such as protons or carbon ions exploit this knowledge to optimize the beam energies to maximize the energy deposited in the tumor.

Figure 4.4 is a sketch of the energy loss features of heavy ions and photons. Photon energy loss peaks at or near the surface, gradually

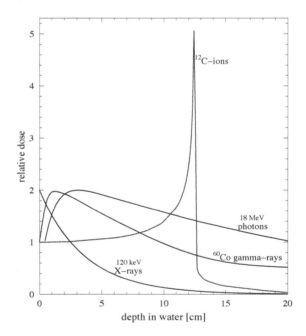

FIGURE 4.4: Dose (energy deposited per unit mass of medium) versus the depth in a water medium for dental (120 keV) X-rays, radiation therapy photons (^{60}Co γ-rays), γ-rays from an electron accelerator (18 MeV) and ^{12}C singly charged ions. (Taken from G. Kraft et al., "First Patient's Treatment at GSI Heavy-Ion Beams" EPAC 1998, p. 212.)

reducing in deposited energy as the depth is increased, spreading the dose deep inside the medium. The charged particles have a finite range and most of the energy is deposited near the end of that range. This feature of heavy ions is favorable for radiation therapies as it reduces the dose to surrounding, non-targeted tissues.

4.5 Range of α Particles

The straggling phenomenon and the Bragg peak are illustrated in simple experiments carried out as student practicums.

Figure 4.5 shows the results of measurements of alpha ranges in an air medium for 5.49 MeV alpha particles from the decay of ^{241}Am. ^{241}Am's half-life of 432.6 years is very handy and inexpensive since the source activity remains nearly constant for several decades. We put the alpha source and a detector in a vacuum chamber separated by some fixed distance. We increase the air pressure in steps. For each pressure setting, we measure the intensities and energies of the alphas reaching the detector per unit time. As we know the energy of emitted alphas (5.49 MeV), we know the energy loss at each setting.

Figure 4.5a shows the number of alphas being detected as we increase air pressure in the chamber, effectively increasing air thickness. Up to about 3 cm thickness, the same number of alphas reach the detector per unit time ($\sim 15 \pm 1$ counts per second). For larger thicknesses, the count rate decreases first slowly, then steeply to about 2 counts/second at 3.5 cm and becomes zero at about 3.7 cm. It is concluded that all alpha particles of 5.49 MeV traverse distances up to 3 cm, with decreasing kinetic energies. Then some of them are stopped between 3 and 3.5 cm. Very few make it to 3.8 cm. Thus the range of alpha particles has some finite spread due to straggling with a mean range of 3.5 cm.

Figure 4.5b is a plot of the kinetic energy of alpha particles as they travel the increasing thickness of the medium. As is seen there, with

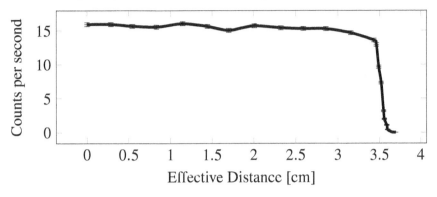

(a) Alpha count rate versus effective distance in air

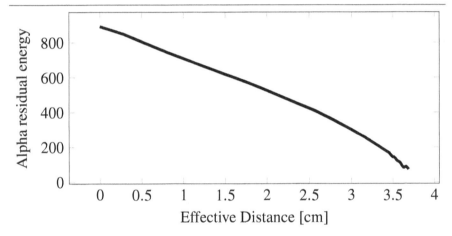

(b) Alpha residual energy in arbitrary units versus effective distance in air

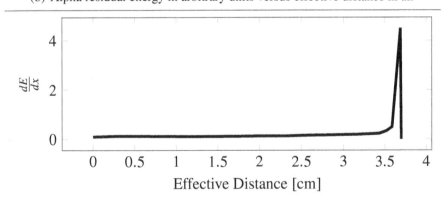

(c) Bragg Peak, energy loss per unit distance versus distance from source

FIGURE 4.5: Passage of 5.48 MeV α-particles of ^{241}Am decay through air medium of various thickness, normalized to standard temperature and pressure (STP).

each increasing distance, alpha particles have less energy, that is, they continually lose energy. We will contrast this with photon propagation, where photon intensities decrease exponentially with an increase in thickness, but the energy of photons reaching the detector is constant.

Figure 4.5c shows energy loss $-(dE/dx)$ of alphas as measured in this experiment. The energy loss is small at small thicknesses, for which alphas have high energies. In deeper layers of the medium, where alphas have lower energy, the energy loss is large, resulting in a Bragg peak at the end. Just before the peak point, the energy loss shows wild fluctuations. While we do not know what exactly is happening, we may speculate that it may be due to an exchange of energy resulting in a gain and loss of energy by the alpha particles. However, the presence of a Bragg peak toward the end of journey of alpha particles is unmistakable. This feature is exploited in radiation therapy using ion beams.

4.6 Range of Particles in Mixtures and Compounds

So far we have dealt with pure elements as media for most practical applications. We now will expand beyond the ideal situation to deal with mixtures and chemical compounds. Bragg provided a simple and profound formula which proved to account for a wide range of species and energies of radiations. For inhomogeneous mixtures of media, the most obvious approach is to sum over the individual contributions of media in the sequence they appear. As the densities of different constituents of a mixture are likely to be distinct, it is necessary to work with mass stopping power.

For $i = 1, 2, ..., n$ material media, each with density ρ_i and thickness dx_i, the mass stopping power is given by Bragg's additivity rule:

$$\frac{dE}{\rho dx} = \sum_i \omega_i \frac{dE}{\rho_i dx_i} \qquad (4.13)$$

For compounds:

$$\frac{dE}{\rho dx} = \frac{1}{\rho} \sum \omega_i \frac{dE}{dx_i} \qquad (4.14)$$

On the right hand side, ω_i is the fraction by weight of the ith element and ρ_i is its density. For compounds, ρ_i is same for all constituents and it can be safely taken outside the summation sign. Thus, in the case of compounds, Bragg's rule applies equally for mass stopping power and linear stopping power. In mixtures, this rule is only for mass stopping power.

Bragg's additivity rule does not take into account the electron binding effects in chemical substances. Over the decades, several researchers investigated the limitations of Bragg's rule. There is general agreement that Bragg's rule is of limited applicability at very low energies and it can be about 50% off from the measurements, especially near energies where electron binding effects are significant.

In our discussions, we briefly touched upon the multiple scattering effects of a particle in the medium, straggling due to energy gains and losses by the particles. This is basically an effect that as particles slow down to very low energies, they are likely to lose or gain energies from the interacting partners. These phenomena cause particles to linger along their path or deviate from their main paths. Generally, the net result is ranges longer than what we expect from theory. Those discussions are beyond our scope.

4.7 Energy Loss of Electrons by Ionization

Electrons, while they ionize media, are interacting with atomic electrons which are their twins. Thus, their interactions with electrons of material media are different from those of heavier particles. In each collision an electron can lose up to 50% of its kinetic energy. As electrons slow down, collisions with atomic electrons in the medium can result in much larger deflections, making them randomly diverted from their path. Also, being of low mass, they acquire speeds comparable to that of light at very low energies and they can lose energy by emitting electromagnetic radiation. The overall energy loss of electrons is a combination of ionization of the medium and emission of radiation. At a given energy, the relative contributions of these two mechanisms depend on electron energy and the medium's properties. A quantitative discussion of the range and energy losses of electrons and photons will be in the next chapter.

4.8   Endnotes

Footnote 3

The kinetic energy of a particle is $KE = (\gamma - 1)mc^2$. This equation includes Newton's equation, $KE = \frac{1}{2}mv^2$ as a low speed limit, as below:

$$(\gamma - 1) = \frac{1}{\sqrt{1-\beta^2}} - 1 \approx 1 + \frac{\beta^2}{2} - 1 = \frac{\beta^2}{2} \text{ for } \beta << 1 \qquad (4.15)$$

$$\therefore KE = (\gamma - 1)mc^2 = \frac{\beta^2}{2}mc^2 = \frac{1}{2}mv^2 \text{ since } \beta = \frac{v}{c} \qquad (4.16)$$

Footnote 4

We call the amount of energy loss per unit distance traveled the stopping power of the medium. This expression is correct for small (infinitesimal) distances (dx) because this equation treats β, the speed of particle, as a constant, while β decreases as the particle loses energy in the medium. For finite distances, we either do the integration or summation over changing β values.

Footnote 5

For those among you who would enjoy an analytic expression, the Bethe formula is written as

$$-\left(\frac{dE}{dx}\right) = \left(\frac{4\pi N_A \alpha (\hbar c)^2}{m_e}\right)\left(\frac{\rho Z}{A}z^2\right)$$

$$\times \frac{1}{\beta^2}\left[\ln\left(\frac{2(\beta\gamma)^2 m_e}{I}\right) - \beta^2 - \frac{\delta}{2} - U\right]$$

In this equation I is the ionization potential. The ionization potential is approximately proportional to the atomic number Z of the atom. Thus $\ln(Z)$ is used to replace $\ln(I)$ in Equation 4.1. Dividing the above equation throughout by ρ, we can define mass stopping power as

$$-\frac{1}{\rho}\frac{dE}{dx} = \left(\frac{4\pi N_A \alpha (\hbar c)^2}{m_e}\right)\left(\frac{Z}{A}z^2\right)$$

$$\times \frac{1}{\beta^2}\left[\ln\left(\frac{2(\beta\gamma)^2 m_e}{I}\right) - \beta^2 - \frac{\delta}{2} - U\right]$$

or

$$-\frac{dE}{d(\rho x)} = \left(\frac{4\pi N_A \alpha (\hbar c)^2}{m_e}\right)\left(\frac{Z}{A}z^2\right)$$

$$\times \frac{1}{\beta^2}\left[\ln\left(\frac{2(\beta\gamma)^2 m_e}{I}\right) - \beta^2 - \frac{\delta}{2} - U\right]$$

The term $\left(\frac{4\pi N_A \alpha (\hbar c)^2}{m_e}\right)$ is a product of fundamental constants: Avogadro's number (N_A), a fine structure constant (α), Planck's constant (h), the speed of light (c) and the mass of an electron (m_e). It is a constant number for all media and particle combinations. Numerically, it is equal to

$$\left(\frac{4\pi N_A \alpha (\hbar c)^2}{m_e}\right) = 307.3 \times 10^{23}\,[\text{MeV F}^2/\text{Mol}]$$

$$= 30.73\,[\text{eV m}^2/\text{Mol}]$$

$$= 0.3073\,[\text{MeV cm}^2/\text{Mol}]$$

or 30.73 [eV m^2/Mol] or 0.31 [MeV m^2 /kMol] or 3.1 [MeV cm^2/Mol].

The alternate ways of expressing the quantity allow one to see easily the number relevant for nuclear dimension (F) or SI units (meter and k-Mol) or cgs units (cm and Mol).

5

Interactions of Photons and Electrons in Matter

5.1 Introduction

The interactions of photons and electrons with matter are intertwined. The interaction of photons with matter is associated with the release of electrons and the generation of electron-positron pairs. The interaction of electrons with matter is of two types. Similar to other charged particles, they undergo continuous slowing down due to collisions with atomic electrons in material media. In addition, they also lose energy by emitting radiation, which in turn propagates in its characteristic ways.

We begin with photon interactions in a medium. First and foremost, we recognize that it is a statistical process in a real sense. Charged particles pass through a medium by leaving ionization trails along the path. For all practical purposes, it can be treated as a continuous deterministic process though, at a very microscopic level, it is governed by quantum mechanical statistical phenomena. On the other hand, a photon may travel an indeterminate distance without any interaction. When it interacts, it is removed from the beam, resulting in a loss of intensity of the beam.

We thus refer to the attenuation of the intensity of a photon beam due to interactions within a medium, rather than energy loss. The probability that a photon interacts in a length interval dx of a medium is independent of the distance it traveled before arriving at the region of interest. This feature suggests an exponential decrease of the photon beam intensity. We may note the similarity with radioactive decay. The decay probability is independent of the past history of a radioactive atom. We cannot talk about a range of distance for photons as we do for charged particle propagation.

5.2   Radiation Length

We define radiation length (X_0) as a characteristic distance denoting the effectiveness of a medium in attenuating photons as they propagate in it. Quantitatively, radiation length is the distance in a medium at which the photon intensity reduces to $1/e$ ($\approx 37\%$) of its initial value.

That is,

$$I(X_0) = \frac{I_0}{e} \approx 0.37 \, I_0 \tag{5.1}$$

The exponential decrease of intensity suggests that the relation for intensity at any arbitrary distance x in the medium

$$I(x) = I_0 \times e^{-\mu x} \tag{5.2}$$

where $\mu = 1/x_0$, a proportionality constant, known as the attenuation coefficient, is a property of the medium.[1]

Clearly, the radiation length has dimensions of length and the attenuation coefficient is of inverse length.

As is common practice, we may define the mass attenuation coefficient (μ_ρ) and the corresponding radiation length (x_0) in units of g cm^{-2} and cm^{2}g^{-1}, respectively.

We define the mass attenuation coefficient (μ_ρ) as $\mu_\rho = \mu/\rho$. In these units

$$I(\rho x) = I_0 e^{-\mu_\rho \cdot \rho x} \tag{5.3}$$

and ρx is the medium thickness in units of g/cm^2.

The definition of mass attenuation coefficients and the presenting radiation length and thickness of a medium in units of g/cm^2 and cm^2/g have the same advantage as in the treatment of charged particles' passage in media, i.e., those parameters do not depend on the physical

---

[1] An astute reader notices the analogy of this equation to that of the radioactive disintegration law. Here μ plays the same role as λ in decay laws. Please refer to Appendix A.

state of the medium. For practical purposes, knowledge of the attenuation coefficient or equivalently the radiation length of a medium for the propagation of photons is all that is required.

The measurement of attenuation coefficients is straightforward.

If we measure intensity variations as we change the thickness (x) of the material, the attenuation coefficient can be deduced as the slope of the straight line plot of $\ln\left[I_0/I_{(x)}\right]$ versus x.

Example 5.1

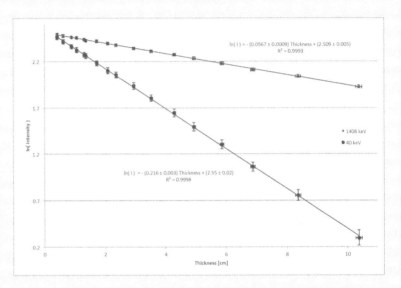

FIGURE 5.1: The natural logarithm of photon intensities versus the thickness of a water column. The intensities are normalized to one (logarithm of intensity normalized to zero) for measurement with no water in the container.

Figure 5.1 shows photon attenuation in a water medium for a pair of gamma rays of energies (40 keV and 1408 keV) emitted by a radioactive source ^{152}Eu. The experimental data is in excellent agreement with the hypothesis of exponential decrease in intensities, since $\ln I$ versus thickness is a straight line with a negative slope.

The slope of the 40 keV data is much steeper than that of the 1408 keV data, showing that low energy photons are more strongly attenuated than those of higher energies. This is true in general with the exception of attenuation near the X-ray absorption edges.

From this experimental data, we deduce that in a water medium

$$\mu = 0.216 \pm 0.003 \text{ cm}^{-1}, \mu_\rho = 0.216 \text{ cm}^2\text{g}^{-1}, \text{ for 40 keV photons}$$
$$\mu = 0.0567 \pm 0.001 \text{ cm}^{-1}, \mu_\rho = 0.0567 \text{ cm}^2\text{g}^{-1}, \text{ for 1408 keV photons}$$

Assuming $\rho = 1$ g/cm^3 as the density of the water medium, we may determine the following information:

In passing through a 1 cm column of water, the intensity of 40 keV photons is reduced to

$$I(1 \text{ cm}) = I_0 e^{-0.216} = 0.8 \, I_0$$

and that of 1408 keV photons is reduced to

$$I(1 \text{ cm}) = I_0 e^{-0.057} = 0.95 \, I_0$$

Thus the 40 keV photon intensity is reduced to 80% and that of the photon 1408 keV is reduced to 95% of the initial values. The differences in the change of intensities become significant for increasing thicknesses due to the exponential character of attenuation.

5.3 Photon Energy and Material Dependence of Attenuation Coefficients

At low photon energies (say $E_\gamma < 100$ keV), the attenuation coefficients are sensitive to excitation spectra of atoms and molecules in the medium. This is due to the selective absorption of photons at characteristic excitations of elements and compounds at the atomic and molecular levels. At higher energies, the attenuation coefficients exhibit monotonous trends as photon energies and material media (Z and A) are varied.

Thanks to several decades of experiments and extensive theoretical calculations, we can now easily obtain the gamma attenuation coefficients to sufficient accuracy from websites. The most authoritative site is XCOM.[2] Here we can get the data for elements, mixtures and compounds, etc. Figures 5.2 and 5.3 show the plot of mass attenuation coefficient versus photon energies 1 keV to 100 MeV for carbon ($z = 6$) and lead ($z = 82$) media.

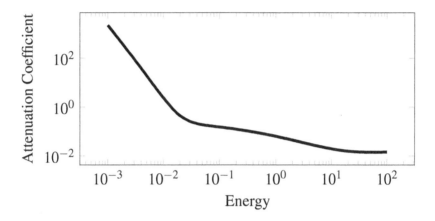

FIGURE 5.2: Total attenuation coefficients of photons in carbon (C) for $E_\gamma \sim 0.001$–100 MeV.

---

[2]The XCOM website: http://www.nist.gov/pml/data/xcom/index.cfm is a culmination of several decades of efforts of researchers at the National Institute of Standards and Technology (USA) and elsewhere.

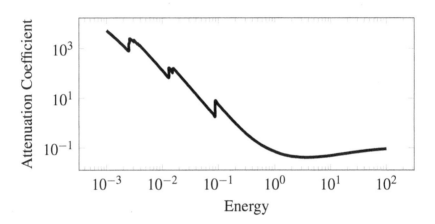

FIGURE 5.3: Photon total mass attenuation coefficients of lead (Pb) for E ~ 0.001–100 MeV.

From the figures, two features become apparent. First, the mass attenuation coefficients of lead are about a factor 10 larger than those of carbon. As lead is almost a factor of six denser ($\rho_{Pb} \sim 11.4, \rho_C \sim 2$), we conclude that, for geometrical considerations, the lead medium is about a factor of 60 more efficient than carbon in attenuating photons. The second feature we note is that attenuation in the lead medium shows a series of discrete jumps up to about 100 keV. We can understand them as follows:

Atoms or molecules exhibit several excited levels. They can be excited from the ground state to some of these levels when incident photons have the correct energies. When they are exposed to external radiation, photons corresponding to the excitation energies of atoms or molecules are selectively absorbed, which is known as resonant absorption. The abrupt jumps in the attenuation coefficients are due to an increase in photon absorption whenever a photon's energy matches these excitations. For lead, one expects enhancement at 2.3, 12.6, 14.77, 72.8, 72.97, 84.94, and 87.32 keV. We can verify this assertion from experiments. If we let a wide spectrum of photons pass through a medium, we will find that photons corresponding to these energies are absorbed more than the neighboring values. Also, if we set a photon detector away from the main beam path, we can detect characteristic lines of atoms. This feature is heavily exploited in scientific research and other applications for identification of material structures by the techniques known as resonance fluorescence.

5.4 Photon Energy Loss Mechanisms

As high energy electrons traveling in high Z media emit photons, the energy loss mechanism of electrons is closely related to that of photons and vice versa. Thus, the interactions of photons, electrons and positrons[3] have a lot of common physics.

---

[3]Positrons are anti-particles of electrons. All properties of positrons are identical to those of electrons except for the electrical charge. Positrons carry an electric charge of $+e$ (1.6×10^{-19} coulombs), of the same magnitude but opposite sign as that of electrons. For particle physics enthusiasts, we say that they also carry a lepton number ($L = -1$) opposite in sign to that of an electron ($L = +1$).

The energy loss of these species is due to electromagnetic interactions. While low energy electrons and positrons ionize media, they emit radiation at higher energies.

In the case of charged particles, we considered elastic scattering and ionization as the main processes of energy loss, and it is a continuous process. While electrons ionize in a continuous manner, radiative processes are dominant for energy losses of high energy electrons, and they are discontinuous.

For radiation, energy loss in a single encounter may be accompanied with or without a photon in the final state. A single encounter with an interacting partner may result in the creation of a charged particle-antiparticle pair which will then continue to lose their energies.

Thus at high energies, the interactions of photons, electrons and positrons are nearly indistinguishable. For most commonly encountered photon energies in nuclear and radiation physics, the particle-antiparticle pairs are electrons and positrons. At energies of several hundred MeV and beyond, we may expect to create pairs of heavier particle-antiparticles such as pairs of pions or heavier pairs as long as energy and momentum conservation conditions are satisfied. We start by discussing photon interactions.

From the early times of studies of cosmic radiation, it was known that cosmic radiation spreads over the entire earth's surface and it is composed of photons and electrons, among other things.

A scientist named Homi Bhabha[4] described this phenomenon as a shower development, currently known as "Bhabha showers." The description is given below.

Consider a photon of very high energy (several MeV or higher). As it interacts with matter on its way (Figure 5.4), it may kick out an electron from an atom or molecule, transferring most of its energy to the electron. Then we have a flying electron and a photon of reduced energy.

Or this photon may create particle-antiparticle pairs. If they are electron-positron pairs of sufficient energy, they may emit radiation

---

[4]Homi J. Bhabha (1909–1966) was a nuclear/particle scientist. The Tata Institute of Fundamental Research and the Bhabha Atomic Research Center of India, now renowned nuclear science and technology research and education institutions of India were his creations.

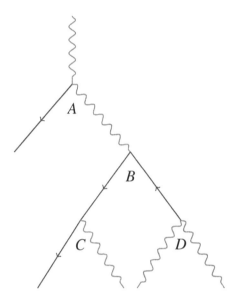

FIGURE 5.4: An illustration of a possible single cascade of events from a single high energy photon. Vertex *A* shows a Compton scattering interaction where the initial photon transfers some momentum to an electron that was bound in a material. At *B*, the photon is lost due to pair production, producing an electron (left of vertex) and positron (right of vertex). Point *C* shows the electron interacting in the medium, emitting a photon due to bremsstrahlung. The electron moves with reduced energy. Finally, *D* shows the positron originally produced by pair production, annihilating with an electron and producing two photons.

known as bremsstrahlung and move with reduced energy in a different direction. In the case of positrons, they may die on electrons in the medium and result in a pair of photons.

You see the point. We may have started with an electron or a photon in the beginning. By the second stage we may have an electron and a photon. If we do not know what we started with, the second stage cannot tell us if we started with a photon or an electron.

5.4.1 Photon Interactions

Interaction mechanisms of photons, while being monotonous functions of the atomic number of the medium they move in, vary with photon energy and atomic number in a distinct way. This feature is exploited to our advantage for each specific application we are interested in.

We list the photon interactions:

Rayleigh scattering This is elastic scattering of photons off atoms or molecules in the medium. It is coherent scattering, where the atom recoils. It involves neither excitation nor ionization of the atom. This process results in a very small loss of the energy of the photon; it is important at very low photon energies and contributes very little to photon energy loss and thus is of little interest in our discussion.

Photoelectric effect A photon is absorbed and an electron is set free from an atom or a molecule. The atom is ionized. We start with an electron bound to an atom or molecule in a material medium and a photon. The interaction results in a free moving electron and no photon. This process effectively removes photons from the beam. The probability for this process increases rapidly with the atomic number of the interacting medium (approximately proportional to Z^4). Also, it plays a significant role for photons of up to about a few hundred keV energy. We recognize that high Z materials such as lead ($Z = 82$) are preferred shielding materials to remove photons and we can understand the use of thin (but heavy) lead sheets in our dentist's office as we are X-rayed.

Compton scattering This phenomenon is elastic scattering of a photon off a free electron. It is incoherent scattering, where an electron leaves the atom and the atom is ionized. There is still a photon after the interaction, albeit of reduced energy. As one can imagine, successive Compton collisions reduce the photons to lower energies where the photoelectric effect dominates and ultimately photons are lost. The Compton process varies roughly proportionally to the atomic number (Z) of the medium. Compton scattering is described as the scattering of a photon off a free electron in a medium. As a material with elements of atomic number Z has Z electrons, the proportionality of Compton scattering to Z is easily understood. This process is dominant for photons of several hundreds of keV to about a few MeV energies.

Pair production A photon is converted into an electron and a positron pair. All of the photon's energy is shared by the particle-antiparticle pair and the recoil electron or the nucleus, participating in the interaction. The photon must have a minimum energy of 1.022 MeV (twice the mass of an electron) to produce particle-antiparticle pairs. This process becomes the dominant mechanism of energy loss of photons for energies greater than about 10 MeV. Also, it varies as Z^2. Thus, higher photon energies and media of heavy materials means a higher probability that photons lose their energy by this process and they disappear. Pair production may have two contributions. One comes from the interactions of photons with atomic electrons and the second is due to interactions with atomic nuclei. It is interesting that the nuclear process dominates for high Z materials.

Photo nuclear reactions This is the interaction of photons with atomic nuclei. The probability for energy loss by elastic or inelastic scattering off nuclei is very small. It is not significant enough to be of concern for the calculation of the energy loss of photons. However, at energies of several MeV, photons can cause dissociation of atomic nuclei, causing emission of nuclear particles. The most common among them is a neutron emission. For all radiation safety and design of nuclear facilities, this process is an important consideration. A major contributing region for this process is giant dipole resonance[5] in nuclei.

It is worth remembering that all processes except Compton scattering and Rayleigh scattering are devoid of photons after interactions.

5.4.2 Photon Interaction Cross Sections

When we send a beam of photons into a medium, the effect is loss of intensity. While individual photons interact by one of the processes mentioned in the preceding section, the overall intensity loss is a cumulative effect of the competing processes, which depend on the probabilities (cross sections) that they occur as molecular, atomic and nuclear interactions. The attenuation coefficient of a medium is a net contribution of all the physical phenomena contributing to the reduction of primary photon intensities.

The attenuation coefficient can be related to the physical processes through atomic cross sections as

$$\mu_\rho = \frac{\sigma_{tot}}{uA} \tag{5.4}$$

where $u = 1.6605 \times 10^{-24}$ g (mass equivalent of the atomic mass unit) and A is the mass number of the medium. Or

$$\sigma_{tot} = A \cdot u \cdot \left(\mu_\rho\right) \tag{5.5}$$

which has dimensions of area. The total cross section (σ_{tot}) is the sum of all possible processes mentioned above, i.e.,

$$\sigma_{tot} = \sigma_{photo} + \sigma_{Rayleigh} + \sigma_{Compton} + \sigma_{pair-nuc} + \sigma_{pair-elec} + \sigma_{photo-nuc} \tag{5.6}$$

The measurement of mass attenuation coefficients provides data to test various physics models of interaction of photons with matter.

As models calculate cross sections of individual processes from nuclear and quantum mechanical theories, one can deduce the partial attenuation coefficients and total attenuation coefficients for specific materials over a finite range of photon energies. Figure 5.5 shows the XCOM model calculation for a lead medium.

---

[5]See Section 5.12.

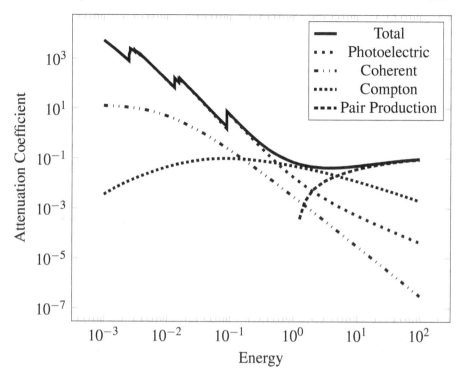

FIGURE 5.5: Total and partial attenuation coefficients in lead for $E_\gamma = 0.001 - 100$ MeV.

At very low energies ($E_\gamma < 100$ keV), the photoelectric effect is the dominant process. For 100 keV–1 MeV both the photoelectric effect and Compton scattering are competing processes. Compton scattering is the dominant phenomenon of attenuation for up to about 10 MeV. For higher energies, pair production is almost the sole phenomenon. Over a narrow region of energies ($E_\gamma = 10 - 20$ MeV) nuclear absorption plays a major role (not shown in the figure).

5.5 Bragg's Additive Rule

As is to be expected, the materials we use in laboratories are rarely pure atoms of a single species. The mass attenuation coefficient of a mixture or compound can be easily calculated from Bragg's additivity expression:

$$\mu_\rho = \sum_i \omega_i \left(\mu_\rho\right)_i \qquad (5.7)$$

where ω_i is the fraction by weight of the ith atomic constituent and $\left(\mu_\rho\right)_i$ is the corresponding mass attenuation coefficient. It is important to recognize that Bragg's rule is identical to the definition used without any change in the case of charged particle interactions. This should not be surprising at all.

For example, water molecules are composed of hydrogen and oxygen in the ratio 2:16 by weight. So we find the atomic mass attenuation coefficients of those atoms from the data tables and deduce the corresponding value for water.

As in the case of interactions of charged particles, Bragg's law is a good approximation for a wide range of energies. However, it does not work well for low energy photons near the resonance energies of compounds. The reason is that the excitation energies of molecules are not identical to those of the constituent atoms.

Example 5.2

From the XCOM database, we read the following data of attenuation coefficients for 1408 keV photons.

Energy of photons: 1408 keV

$\left(\mu_\rho\right)$ in H_2 medium: 0.1062 cm²/g

$\left(\mu_\rho\right)$ in O_2 medium: 0.0536 cm²/g

In water (H_2O), the ratio of the weight of oxygen and hydrogen is 16:2. Or 18 g of water consists of 2 g of hydrogen and 16 g of oxygen.

Then from Bragg's additive rule:

$$\left(\mu_\rho\right)_{\text{water}} = \frac{2}{18}\left(\mu_\rho\right)_{\text{hydrogen}} + \frac{16}{18}\left(\mu_\rho\right)_{\text{oxygen}}$$

$$= \frac{1}{9} \times 0.1062 + \frac{8}{9} \times 0.0536$$

$$= 0.059 \text{ cm}^2/\text{g}$$

In Example 5.1 we found

$$\left(\mu_\rho\right)_{\text{water}} = 0.057 \pm 0.001 \text{ cm}^2/\text{g}$$

Bragg's estimate and the experimental data agree to within 2%.

5.6   Electron Interactions

Electrons are light charged particles and they become relativistic at very low energies; they are minimum ionizing for $\beta\gamma \sim 3$, i.e., $E_e \sim 1.5$ MeV.

While heavy charged particles travel nearly in straight lines, gradually losing energy to come to rest at a distance well defined as the range of the particle, electrons are easily deflected from their path. The result is that they follow zig-zag paths extended to long distances, especially near the end of their journey.

For electrons, energy loss by radiation (known as bremsstrahlung) becomes dominant for energies greater than a critical energy.[6]

$$E_c \sim \frac{800}{(Z+1.2)} \text{ MeV} \tag{5.8}$$

Note that the constant term (1.2) in the denominator is very important for light nuclei, such as for hydrogen ($Z = 1$), while it may be neglected

[6]See Section 5.12.

for high Z materials.

$$E_c \sim \frac{800}{2.2} \approx 400 \text{ MeV for hydrogen, } Z = 1$$

$$E_c \sim \frac{800}{83.2} \approx 10 \text{ MeV for lead, } Z = 82$$

Thus, electrons radiate readily in high Z media. In lead material, the radiation is significant for energies about 10 MeV and higher, while it is not important up to about 400 MeV for a hydrogen medium. This is the reason X-ray machines and other photon beam facilities employ high Z materials such as lead and tungsten to produce radiation.

5.7   Bremsstrahlung

This radiation is produced when a charged particle traverses a medium. The charged particles experience an accelerating/decelerating force and the result is radiation. The cross section for emission of these photons is given by

$$\frac{d\sigma}{dE_{\text{photon}}} \approx \frac{10\alpha Z^2}{E_{\text{photon}}} r_e^2 \frac{\gamma}{(\gamma - 1)}$$

$$\approx \frac{5.8\, Z^2}{E_{\text{photon}}} \frac{\gamma}{\gamma - 1} \left[\frac{\text{mb}}{\text{MeV}} \right] \text{ for } E_e > E_\gamma \qquad (5.9)$$

Where the fine structure constant is

$$\alpha = \frac{e^2}{\hbar c} = \frac{1}{137} \qquad (5.10)$$

The classical electron radius is

$$r_e = \frac{e^2}{mc^2} = 2.813 \text{ fm} \qquad (5.11)$$

γ is the Lorentz factor

$$\gamma = \frac{E_e}{m_e c^2} \qquad (5.12)$$

A look at this expression shows a few features. Besides the constant, for a fixed energy, the probability of emitting a bremsstrahlung photon decreases inversely as the energy of the photon ($\propto 1/E_\gamma$). It varies as the square of the atomic number of the medium.

The spatial distribution of radiation $I(\theta)$, where θ is the angle with respect to the direction of propagation of the electron, in radians,[7] is given as

$$I(\theta) = \frac{\gamma^4 \left(1 + \theta^4 \gamma^4\right)}{\left(1 + \theta^2 \gamma^2\right)^4} = \frac{\frac{1}{\gamma^4} + \theta^4}{\left(\frac{1}{\gamma^2} + \theta^2\right)^4} \tag{5.13}$$

which, for large γ, becomes $I(\theta) \propto 1/\theta^4$. Thus, bremsstrahlung is confined to very small angles, moving in the forward direction.

The forward cone is, roughly speaking, confined to an angle of $\theta = m_e/E_e$ radians. For 1 GeV electrons

$$\gamma = \frac{0.511}{10^3} \approx 0.5 \times 10^{-3}$$

This feature can be an advantage for many experiments and it can be blinding to some detectors, since all the photon flux is confined in a narrow cone. It depends on what you want to do. We may summarize electron interactions in material media as follows:

1. High energy electrons in high Z media are more prone to energy loss by radiation. An electron is more likely to lose its energy by emitting low energy photons, which means each electron is a source of several (a multitude) of photons.

2. At each stage, an electron is deflected away from its path, albeit by a small angle. A result could be that an electron may be quite far from its original path toward the end of its journey. This is different from heavier particles, which move nearly along a straight line path.

High Z materials cause radiation more readily than lighter ones. Note the energy dependence ($\propto 1/E_\gamma$). The cross section goes almost to infinity for very low energies. Thus most of the flux is concentrated in this region, while the spectrum extends all the way up to an electron's kinetic energy.

---

[7] 1 radian = 57.29 degrees.

5.8　Radiation Length

Though we are not able to ascertain when a photon interacts in a medium, we can still estimate the average energy loss of a beam of photons. One defines radiation length[8] (X_0). It is the distance over which an electron or a high energy photon loses all but 1/e of its energy through bremsstrahlung or the pair production process.

It is approximately given by $X_0 = \frac{716}{Z^2}$ g/cm^2.

To a very good precision, it is written as

$$X_0 = \frac{716.4A}{Z(Z+1)\ln\left(\frac{287}{\sqrt{Z}}\right)}\,\text{g}\cdot\text{cm}^{-2} \tag{5.14}$$

We may thus express the medium length (l) in units of radiation length (l/x_0). The power of the beam, after it traverses a distance of l/x_0, is given by

$$P(l) = P(0)e^{\frac{-l}{X_0}} \tag{5.15}$$

where $P(0)$ is the input power $P(l)$ is the power after the radiation propagates through a distance l.

Table 5.1 lists the radiation length, density and critical energy of a few materials of interest for radiation detection and shielding purposes. Also listed are the densities of the materials. The large variation of physical sizes of material media is easy to recognize.

It is easy to recognize that a liquid hydrogen column of 1 cm is as effective as a gas hydrogen column of about 10 meters.

---

[8]Above, we already defined radiation length as $X_0 = 1/\mu$. For low energies, this parameter increases with increasing energies, since energy loss and attenuation coefficient decrease drastically before they become nearly constant at high energies. As remarked above, pair production is the principal interaction mechanism of radiation.

TABLE 5.1: Data from the Particle Data Group.

Material	Density $(g\ cm^{-3})$	Radiation length $(X_0\ g\ cm^{-2})$	Critical energy (MeV)
hydrogen gas	8.38×10^{-5}	63.04	344.8
hydrogen liquid	0.0708	63.04	278
silicon	2.33	21.82	40.2
germanium	5.32	12.25	18.1
zirconium	6.51	10.2	14.7
sodium iodide	3.67	9.49	13.3
lead	11.4	6.37	7.4
uranium	19.0	6.0	6.6

5.9 Cherenkov Radiation

Cherenkov radiation is emitted when a charged particle traverses a medium with a velocity greater than that of the speed of light in that medium.

If n is the refractive index of a medium, we know that $v_{light} = \frac{c}{n}$ where c is the speed of light in a vacuum.

Thus, Cherenkov radiation is emitted if $v > v_{light} = \frac{c}{n}$ or $\beta n > 1$.

Cherenkov light is emitted as a cone of angle Θ_c, relative to the direction of the charged particle. The cone angle is given by[9]

$$\cos \Theta_c = \frac{1}{n\beta} = \frac{1}{n} \left[1 + \frac{m^2}{p^2} \right]^{1/2} \qquad (5.16)$$

Since the cosine of an angle is less than or equal to 1

$$\frac{1}{n\beta} \leq 1 \text{ or } \beta \geq \frac{1}{n}$$

Thus $\beta_{min} = \frac{1}{n}$ for the emission of Cherenkov radiation. We see that the corresponding threshold momenta and total energies are

---

[9]Here and below we drop c^2 for mass-energy terms. They are implicit and dropping them should not affect the conclusions, as long as we use energy units.

$$\text{Threshold Momentum: } P_{\min} = \gamma\beta m \tag{5.17}$$

$$\text{Threshold Total Energy: } E_{\min} = \gamma m \tag{5.18}$$

From these equations we note the following: Lighter particles emit radiations at smaller energies than heavier ones. Clearly electrons, being the lightest charged particles, emit radiations at very low energies. The emission angle for a particle of mass m and momentum p is given by

$$\theta_c = \arccos\left[\frac{1}{n} \times \left(1 + \frac{m^2}{p^2}\right)^{\frac{1}{2}}\right] \tag{5.19}$$

or $m = p[n^2 \cos^2(\theta_c) - 1]^{\frac{1}{2}}$.

In a given medium, larger masses emit radiation at smaller angles. We can then use Cherenkov radiation information in two ways.

Threshold method If a particle of known momentum p does not emit radiation, then its mass $m > p(n^2 - 1)^{\frac{1}{2}}$ in energy units.

Differential method If Cherenkov radiation is emitted at an angle θ_c, the mass of a particle is given as

$$m = p\left(n^2 \cos^2(\theta_c) - 1\right)^{\frac{1}{2}} \tag{5.20}$$

If the momentum p is in GeV/c or MeV/c, then the corresponding mass of particles is in GeV/c^2 or MeV/c^2, respectively.

Example 5.3

In a water medium of $n = 1.3$,

$$\beta_{min} = 0.77; \gamma_{min} = 1.564,$$

and $\text{K.E.}_{min} = (\gamma_{min} - 1)mc^2 = 0.564mc^2$ for the emission of Cherenkov radiation by a particle of mass m. Thus in water the threshold parameters for emission of Cherenkov radiation are

Particle	Mass [MeV/c^2]	K.E.$_{min}$ [MeV]
electron	0.511	0.29
pion	139.5	78.7
kaon	493.8	278.5
proton	938.3	529.2

Clearly, electrons being the lightest particles, they emit Cherenkov radiation even at very low energies.

For example, the refractive index of water $n = 1.3$ suggests $\beta_{min} = 1/n = 1/1.13 = 0.77$, $\gamma = 1.6$.

Thus, the threshold energies are 0.8 MeV for electrons and about 1.6 GeV for protons. At water pool nuclear reactors, we see a blue light spread across the water column. It is the Cherenkov light emitted by the electrons produced in neutron decays and other radioactive isotopes produced in nuclear fission.

Example 5.4

Charged particles of momentum 1 GeV/c traverse a medium of refractive index $n = 1.1$. What are the Cherenkov emission angles for various charged particles?

We have

$$\theta_c = \arccos\left[\frac{1}{n} \times \left(1 + \frac{m^2}{p^2}\right)^{\frac{1}{2}}\right]$$

$$p = 1 \text{ GeV}/c = 1000 \text{ MeV}/c$$

$$\theta_c = \arccos\left[\frac{1}{1.1} \times \left(1 + \frac{m^2}{10^6}\right)^{\frac{1}{2}}\right]$$

$$= \arccos\left[0.91\left(1 + m^2 \times 10^{-6}\right)^{\frac{1}{2}}\right]$$

with the mass of the particles in MeV/c^2 units.

Particle	Mass	θ_c in medium of $n = 1.1$
electron	0.511	24.5°
pion	139.5	21.9°

We can calculate the minimum momentum at which kaons and protons radiate as

$$p_{min} = \frac{m}{\sqrt{n^2 - 1}}$$

for a kaon $m = 493.8$ MeV/c^2; $p_{min} = \dfrac{493.8}{\sqrt{(1.1)^2 - 1}} = 1077$ MeV/c

for a proton $m = 938.3$ MeV/c^2; $p_{min} = \dfrac{938.3}{\sqrt{(1.1)^2 - 1}} = 2047$ MeV/c

Kaons are just near the threshold for radiations while protons should be of at least twice as much momentum for them to radiate.

Cherenkov light detection is extensively used as a technique to discriminate between and identify different species of charged particles.

5.10 Transition Radiation

While Cherenkov radiation is emitted in a homogeneous medium, transition radiation is emitted when a particle traverses the boundary of two media with different dielectric constants. A dielectric medium is a poor conductor of electricity. In many ways, the term "dielectric" is synonymous with "insulator." In general, the speed of light in a medium (v_{light}) is given by

$$v_{light} = (\mu \varepsilon)^{\frac{1}{2}} \tag{5.21}$$

where the permeability (μ) depends on the magnetic properties of the medium and the permittivity (ε) relates to electrical properties.

For non-magnetic materials, the dielectric constant (k) is a direct measure of the speed of light in a medium and the dielectric constant and refractive index (*n*) of a medium are related by $k = n^2$.

Thus, when a charged particle traverses a boundary of two media, the influence of the electrons in the medium and the charged particle on each other results in the emission of radiation by the charged particle. In some ways, we can imagine this is similar to the refraction of light (change in the direction of motion) as they cross boundaries of materials of different refractive indices. It is due to the fact that a medium exerts force on radiation and charged particles. The charged particles, when subjected to an external force, emit radiations. We thus have an intuitive understanding of transition radiation.

Detailed theory[10] shows that the energy of emitted radiation across the boundary is proportional to γ, the Lorentz factor of the charged particle.

We must contrast γ at the boundary of two media with β, the dependence on the speed of particles for the emission of Cherenkov radiation in a homogeneous medium.

Quantitatively, the maximum energy of transition radiation is ($E_{\text{radiation}}^{\max}$) at a single boundary is given by

$$E_{\text{radiation}}^{\max} = \gamma \times \hbar\omega_p \tag{5.22}$$

where γ is the Lorentz factor of the charged particle and $\hbar\omega_p$ is the energy associated with the electron plasma frequency of the medium from which the particle enters the boundary.

Electron plasma frequency is the parameter used to describe electron density oscillations and the energy associated with it and is given by

$$\hbar\omega_p \approx 30 \text{ eV}$$

for commonly encountered media such as plastics.

---

[10]A detailed, comprehensive theory can be found in several articles. See, for example, *Practical theory of multilayered transition radiation detector* by X. Artu, G. B. Yodh and G. Mennessier, in Physical Review D, Volume 12, pages 1289–1306, (1975).

Example 5.5

We can compare the maximum energies of transition radiation for particles of 1 GeV/c momentum of Example 5.4

Particle	β	γ	$n_{max} = \frac{1}{\beta}$	$E_{transition}^{max} =$ $30 \times \gamma$ [eV]
electron	0.99999	2000	1.0	60,000
pion	0.99047	7.2	1.0096	217
kaon	0.89692	2.2	1.1149	67
proton	0.72935	1.4	1.3710	43

Clearly, electrons most readily emit high energy photons as transition radiation, i.e., 60 keV compared to about 217 eV by pions, the next lightest particle.

At this momentum, it is almost impossible to use time of flight technique or Cherenkov radiation to distinguish electrons and pions from each other. Transition radiation measurements are a good alternative to this end.

5.10.1 Synchrotron Radiation

The term synchrotron radiation refers to the radiation emitted by electrons as they pass through a magnetic field. Though this term was coined for the light observed in modern electron synchrotrons, a type of particle accelerator invented in the 20th century, the radiation from astronomical sources such as the Crab Nebula[11] is described as synchrotron radiation.

As we will see in the chapter on radiation sources, circular accelerators were developed to enhance the efficiency to reach higher and higher energies of charged particles. To this end, the particles should be bent repeatedly to retrace their path for further acceleration. A basic arrangement consists of an electromagnet, arranged as a dipole acting as north and south poles, very similar to the permanent magnets we are all familiar with. Of course, the systems are more sophisticated,

---

[11]The Crab Nebula is known as a supernova remnant first observed by Chinese astronomers in the year 1054 A.D.

well-designed and well-built complex magnet structures to ensure that the particles stay their course for many revolutions. Ironically, a relativistic particle bent this way emits radiation and loses energy, thus undermining the very purpose of magnetic field arrangements. This loss of energy to radiation is significant for electrons and it must be compensated to reach higher energies. Understandably, the first reaction of accelerator builders was that this phenomenon was a headache. Soon, the potential use of this radiation as a tool for applications in diverse science, technology and industry became apparent. Several electron synchrotron facilities, simultaneously serving multiple users, have been built in the last few decades. Dubbed "light sources," they are widely used around the world.[12]

Intuitively, we can understand the physics of synchrotron radiation as follows: Every charged particle has a radiation field along with it. The field energies vary with particle energies. As long as a particle moves in a straight line, the radiation and charged particle move together. An external field can bend a charged particle but not its radiation. The radiation field tends to move in a straight line but the charged particle tends to keep it to itself. The net result is a finite loss of energy to the radiation at each bend.

While the mathematics of synchrotron radiation emission is quite involved, the final results are not difficult to understand. We discuss the salient features here. As we emphasized several times, a charged particle of fixed momentum p is bent in a magnetic field B to travel on a path of the radius of curvature ρ.

It is thus clear that the radiation loss or the energy and intensity spectrum of the emitted radiation depends on the particle momentum p and the applied magnetic field B. We should then adjust these two parameters to achieve the synchrotron light spectrum that we desire for the application at hand.

For an electron bending at radius ρ, the synchrotron light is confined to a narrow cone along its initial path. For a given magnetic field arrangement of electrons of fixed energy, we define a critical frequency

---

[12]The website http://www.lightsources.org/ is a good resource where one can also find links to light sources around the world.

ω_c of synchrotron radiation as

$$\omega_c = \frac{3}{2}\frac{c}{\rho}\gamma^3 \tag{5.23}$$

where c is the speed of light (3×10^8 m/s) and

$$\gamma = \frac{E_e}{m_e c^2} = \frac{E_e\,[\text{GeV}]}{0.511 \times 10^{-3}} = 1,957 \times E_e\,[\text{GeV}] \tag{5.24}$$

where $E_e\,[\text{GeV}]$ is the energy of electrons in GeV.

Also, we can identify a critical angle for the emission of radiation of frequency ω as

$$\theta_c(\omega) = \frac{1}{\gamma} \times \left(\frac{\omega_c}{\omega}\right)^{\frac{1}{3}} \tag{5.25}$$

Thus, for each arrangement of electron energies parameterized by γ, magnetic field B and the bending radius ρ, there is a critical frequency. The synchrotron radiation above this frequency is negligible. Also, the radiation is confined to a narrow cone. The critical angle, as specified above, is unique to each frequency and the radiation of that frequency is confined to a narrow cone around the critical angle. Clearly, low energy radiation (higher wavelength of smaller frequency) is more spread out than higher energy radiation.

Example 5.6

The Canadian Light Source (CLS) in Saskatoon (Canada) accelerates electrons to 2.9 GeV.

Thus, $\gamma(\text{CLS}) = 1957 \times 2.9 = 5088$.

The Super Photon Ring - 8 (SPring8) in Harima (Japan) accelerates electrons to 8 GeV.

$\gamma(\text{SPring8}) = 1957 \times 8 = 15,656$.

For the same bending radius ρ at the CLS and SPring8, the ratio of critical frequencies at these facilities is

$$\frac{\omega_c(\text{SPring8})}{\omega_c(\text{CLS})} = \left(\frac{\gamma(\text{SPring8})}{\gamma(\text{CLS})}\right)^3 = \left(\frac{15656}{5088}\right)^3 = 29.13$$

The critical frequency is nearly a factor of 30 higher than that of the CLS arrangement. We can compare the ratios of critical angles for a specific frequency ω. We find

$$\frac{B(\text{CLS})}{B(\text{SPring8})} = \left[\frac{\theta_c(\omega, \text{SPring8})}{\theta_c(\omega, \text{CLS})}\right]$$

$$= \left(\frac{\gamma(\text{CLS})}{\gamma(\text{SPring8})}\right)\left[\left(\frac{\gamma^3(\text{SPring8})}{\gamma^3(\text{CLS})}\right)\right]^{\frac{1}{3}}$$

$$= 1$$

That is, the synchrotron light cone will be of the same size at both facilities. However, since the critical frequency of the CLS arrangement is about 3% of the value at the SPring8, the intensities of photons at the CLS is limited to those low frequencies with negligible fluxes at higher values.

Rivikin[13] provides a useful approximation to calculate intensities in a bending magnet.

5.10.1.1 Photon Flux

$$Flux = 2.46 \times 10^{13} \times E \times I \times G_1(x) \qquad (5.26)$$

where

$$G_1(x) = x^{\frac{1}{3}} \times 2.11 \times \left[1 - \left(\frac{x}{28.17}\right)^{0.848}\right]^{\frac{1}{0.0513}} \qquad (5.27)$$

and $x = \omega/\omega_c$. Figure 5.6 is the graphical representation of $G_1(x)$.

Here the flux is calculated as the number of photons per second of a given frequency with a spread of 0.1% ($\frac{d\omega}{\omega} = 0.0001$), spreading over one milliradian (< 0.06 degrees) around the critical angle. As before, E is the energy of electrons in GeV and I is the electric current in amperes in the synchrotron.

At any facility, for a bending magnet arrangement, we can easily calculate the intensity profile to a good approximation.

Simple dipole arrangements show monotonous spectrum shapes,

---

[13]Lenny Rivikin, presentation at CERN Accelerator School, Bulgaria (September 2010).

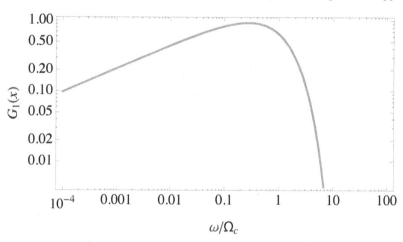

FIGURE 5.6: Plot of $G_1(x)$ versus the ratio of frequency of interest, ω, to the critical frequency, ω_c.

as we indicated above. Scientists and engineers realized that one can custom build large numbers of magnetic structures to influence the intensity profile to suit their needs. These magnet arrangements can be chosen to selectively increase intensities in energy regions of interest. These magnetic structures are called "undulators" and "wigglers." The terms simply refer to the extent to which electrons deviate from their intended path. Undulators keep the deviations small, producing high intensity, nearly monochromatic beams. Wigglers produce high intensity, high energy photon beams that are less monochromatic.

Figure 5.7 is the photon spectrum of the SPring8 facilities. A few characteristics are worth noting. As mentioned above, the radiation from a bending magnet is continuous with a cutoff high frequency. A wiggler increases the intensity of high energy photons by one to two orders of magnitude and cuts off the low energy spectrum (the dashed line portion of the Wiggler curve. An undulator produces spikes of high intensities of specific energies.

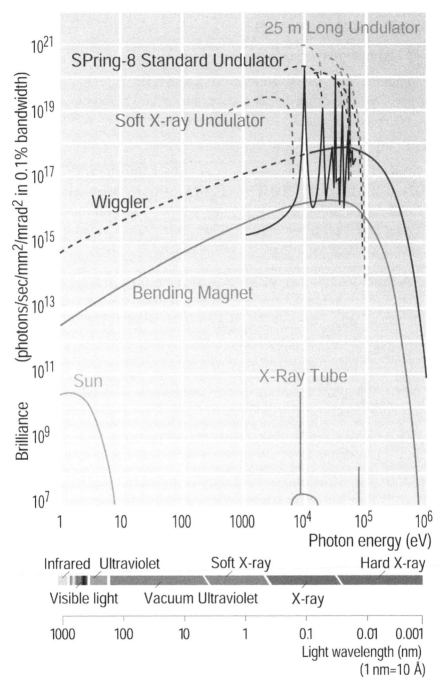

FIGURE 5.7: Spectra of various photon beams produced by the SPring-8 facility. (Photo courtesy of RIKEN, Japan.)

5.11   Questions

1. Photons of 2 MeV and 15 MeV energies pass through a lead medium. Describe the secondary and tertiary radiations that will be generated. What are the physics phenomena responsible for these emissions?

2. Calculate the half-value thickness (in centimeters) of NaI and a lead medium for 100 keV, 1 MeV and 15 MeV photons. The half value thickness is the amount of material needed to reduce the intensities to 50% their initial values. Please visit http://www.nist.gov/pml/data/xcom/index.cfm

3. A beam of 200 keV photons is incident on a 1.4 cm thick sheet of aluminum (density = 2.7 g cm^{-3}) pressed against a 2 mm thick sheet of lead (density = 11.34 g cm^{-3}) behind it. The mass attenuation coefficient is 0.1223 cm^2/g for aluminum and 0.999 cm^2/g for lead.

   (i) What fraction of incident photons enter the lead sheet without interacting with aluminum?

   (ii) What fraction of incident photons penetrate both sheets without interacting?

   (iii) If the direction of the photons changed or we interchange the aluminum and lead so that photons enter the lead first, what fraction of photons enter aluminum without interacting with lead?

4. Photons of 1 MeV energy pass through a column of 10 cm of water in a glass container with 2 mm thick walls.

   The mass attenuation coefficients are 0.0707 cm^2/g and 0.0634 cm^2/g for water and the glass medium, respectively.

   The density of water is 1.0 g/cm^3 and that of glass is 2.23 g/cm^3.

   What fraction of incident photons

   (i) enter the water medium,

    (ii) exit the water medium

    (iii) exit the entire assembly.

5. A bronze absorber (density = 8.79 g cm^{-3}) is made of copper (Cu) and tin (Sn). Its composition is 90% Cu and 10% Sn, by weight. Calculate the linear and mass attenuation coefficients of bronze when it is exposed to 200 keV X-rays. The mass attenuation coefficients of Cu and Sn are 0.15 cm^{2}g^{-1} and 0.31 cm^{2}g^{-1} for photons of 200 keV.

6.   (i) The human body is made up of a mixture of 55% water, 1.9% calcium, 0.9% phosphorous and organic medium. The percentages are by weight.

    Calculate the mass attenuation coefficient of the human body for 146 keV gamma rays, as if it were only due to these materials. The required data is given below:

Medium	Density [g cm^{-3}]	Linear Attenuation μ [cm^{-1}]
water	1.00	0.1581
calcium	1.55	0.2652
phosphorus	1.8	0.2615
average of the human body	1.4	

  (ii) Consider the organic medium to be hemoglobin, whose chemical constituents may be approximated to be of composition Fe N$_4$C$_{12}$H$_{23}$O$_2$. Its density is 1.06 g cm^{-3}. The mass attenuation coefficients of the constituent elements are (in units of g cm^{-2}): ^{56}Fe: 0.2032; ^{15}N: 0.1364; ^{12}C: 0.1340; ^{1}H: 0.2671; ^{16}O: 0.1373.

    Calculate the mass attenuation coefficient of hemoglobin.

  (iii) From the results of (i) and (ii) calculate the mass attenuation coefficient of the human body.

7. A beam of electrons, pions and protons passes through water ($n = 1.33$), plastic ($n = 1.5$) and aerogel ($n = 1.08$). Calculate the threshold momenta at which these particles emit Cherenkov radiation in these media. If the particles are of momenta $p = 500$ MeV/c, in

which of the above media do these particles, if any, emit Cherenkov radiation?

8. From critical energy considerations, estimate the energy at which radiation becomes a dominant process of energy loss of electrons in lead, copper and water.

Visit the website

http://physics.nist.gov/PhysRefData/Star/Text/ESTAR.html

A 300 MeV electron beam passes through these materials. From the website, find the energy losses of these electrons by collision and radiative processes for these materials. Are the data consistent with your expectations of critical energy considerations?

9. The atomic cross sections for 1 MeV photon interactions with carbon and hydrogen are 1.27 and 0.209 barns, respectively. Calculate the linear attenuation coefficient for paraffin. Paraffin is CH_2 of density 0.89 g cm^{-3}.

10. The European Synchrotron Radiation Facility (ESRF), located in Grenoble, France, operates at an electron energy of 6 GeV. Assume the electrons pass through a magnetic field of 0.5 Tesla. For this arrangement, calculate

   (i) the critical frequency

   (ii) $G_1(x)$ function (see text)

   (iii) the photon intensity spectrum for electron beam currents of 100 mA.

5.12 Endnotes

Footnote 5: Nuclear Giant Dipole Resonance

While nuclear structure and most nuclear processes are very specific to each atomic nucleus without predictable Z dependence, they still exhibit one salient feature, common to all nuclei, known as giant resonances. These are excitations in which almost all particles in the nucleus play a collective role. Without worrying about theoretical details, we note that these resonances occur at around $66 \times A^{-1/3}$ MeV, where A is the mass number of atomic nuclei in the medium. For example, they play an important role for about 10–20 MeV photons in a lead medium and at about 15–25 MeV for a lighter medium such as iron. These resonances decay predominantly by emitting neutrons. The cross sections for this process are quite large. This process is a serious safety consideration at all particle acceleration facilities.

Footnote 6

Rigorously, critical energy is defined as the energy at which both the ionization process and radiation contribute nearly equally to the energy loss of charged particles. We use Bethe's energy loss formula for the ionization process and radiation energy loss by bremsstrahlung and equate the two contributions to determine the critical energy. We should get an expression very close to Equation (5.8).

6

Interactions of Neutrons with Matter

6.1 Introduction

This chapter is devoted to the interactions of neutrons with matter. This topic is of interest when we are looking for uses in science and technology or we are concerned with the radiation damage and hazards they pose. Neutron interactions with material media share some features with charged particles and some others with photons, but some aspects of these interactions are unique to them. We can talk about the slowing down of neutrons, i.e., neutrons losing energy by multiple elastic collisions along their path in the medium, in the same way as for charged particles. Their interactions resemble those of charged particles in that the neutron after scattering is the same one we started with except it is of lower energy and is deflected from its original path. Unlike the case of charged particles, which leave ionization trails along their paths, neutrons leave no track. Also, analogous to photon propagation, they may travel long distances without interactions and be absorbed in a single encounter.

Neutrons interact mainly with atomic nuclei over a wide range of energies. This behavior is in contrast to the interaction of charged particles and photons, for which energy loss or absorption is mainly due to interactions with electrons. To a lesser extent, they also interact with atomic/molecular electrons at low energies.[1]

As they are electrically neutral, heavy and are of small size, neutrons can penetrate deep into the interiors of atoms even at low energies, get very close to atomic nuclei and thus be readily absorbed by them. The absorption is an exoergic process. It is generally followed by the emission of photons, sometimes charged particles and in some

---

[1] See Section 6.6.

cases by the well known nuclear fission. Quite often, the interactions produce radioactive nuclei which decay with their typical half-lives and emit radiations.[2]

We found that the energy loss of charged particles or photons in interactions varies monotonously with the atomic number (Z) of the medium. Thus, from the knowledge of atomic number and other physical properties such as density, we could very easily assert which medium is more suitable for a physics application. We could also specify the amount of energy deposits along the path of charged particles and their ranges. In contrast, except for energy loss by elastic scattering, absorption of neutrons is very specific to the medium not only with regard to atomic number but also mass number. While fractional energy loss by elastic scattering varies smoothly with the mass number of the medium, the absorption of neutrons is unique to the nuclear isotope present in the medium. For example, the Q-value for the (n, α) reaction is 2.79 MeV and -6.63 MeV for the ^{10}B and ^{11}B targets, respectively. The positive Q-value and that the cross section of the (n, γ) process on the ^{10}B nucleus is several thousand barns makes this process a very attractive option for the detection of neutrons. The ^{11}B isotope is not useful in this regard as very low energy neutrons will mostly bounce off ^{11}B atoms.

Understandably, neutron interactions with materials play an important role in the choice of media for the slowing down of neutrons in a nuclear reactor or in the design of a neutron detector. We pick materials like cadmium and gadolinium for neutron absorption and ^{3}He or ^{10}B for neutron detection. The types of radiations, their intensities and levels of residual radioactivities are quite sensitive to the interacting nucleus and neutron energies. This happens because each nuclear isotope has several unique properties, such as levels of excitation, quantum mechanical properties, and binding energy relative to neighboring nuclei which determines relative stabilities of nuclear systems.

Figure 6.1 is a plot of neutron total interaction (scattering, absorption, etc.) with aluminum for a wide range of energies from a few micro-electron volts (10^{-6} eV) to about tera electron volts (10^{12} eV), spanning nearly 18 orders of magnitude in energy. The cross section

---

[2]This is the basic physics behind neutron activation to produce radioactive isotopes for industrial use, medical purposes or academic research.

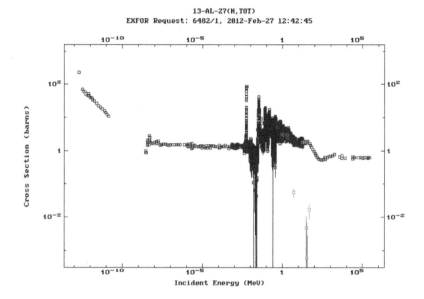

FIGURE 6.1: A graphical presentation of neutron interaction probability in an aluminum medium. Source: NNDC website. [Courtesy of EXFOR Web application (http://www.nndc.bnl.gov/exfor).]

also changes by nearly four orders of magnitude from 100 barns to 0.01 barns or less. The cross section changes in a very conspicuous way. Starting at a very high cross section of about 100 barns for neutrons of 1 μeV energies, it decreases to about 1 barn for 1 meV neutrons and stays nearly constant up to about 10 keV neutrons. From 10 keV to about 1 MeV, the cross section shows wild fluctuations, with several maxima and minima. The maxima are resonance reactions where a neutron is preferentially absorbed to form a compound nucleus. It may then be re-emitted or lost from the medium, resulting in the emission of photons or other charged particles. The location of resonances and the strength of the interaction are features specific to the compound nuclear (neutron + nucleus of the medium) system.

A compound nucleus formation is a reaction in which a projectile is absorbed by the target nucleus to form a composite system. Subsequently, the composite system settles to a final state which solely depends on the properties of the composite compound nucleus.

In the case of a neutron projectile on a target nucleus (A, Z), the compound nucleus is $(A + 1, Z)^*$, where $*$ indicates an excited level of the $(A + 1, Z)$ nucleus.

At much higher energies ($E > 1$ GeV), the cross sections are again nearly constant.

Example 6.1

Neutrons on a ^{10}B nucleus may form a compound nucleus of ^{11}B*, an excited level in ^{11}B.

The mass excess of ^{10}B, ^{11}B and n are

$\Delta(^{10}\text{B})$	12.051MeV
$\Delta(^{11}\text{B})$	8.668 MeV
$\Delta(n)$	8.071 MeV

So the Q-value for

$$^{10}\text{B} + n \rightarrow ^{11}\text{B}^*$$

is

$$\Delta(^{10}\text{B}) + \Delta(n) - \Delta(^{10}\text{B}) = 11.454 \text{ MeV}$$

Thus neutron absorption leads to excited levels at and above 11.454 MeV in ^{11}B. The excited levels settle to final states. In the case of (n, α) reactions, we see this process as

$$^{10}\text{B} + n \rightarrow ^{11}\text{B}^* \rightarrow ^7\text{Li} + \alpha$$

with ^{11}B* as the intermediate, compound nucleus state.

6.1.1   Nomenclature of Neutrons

In the literature, the nomenclature of neutrons of different energies is a hodgepodge choice referring to neutron speeds, the concept of temperature from thermodynamics and physics features of neutron interactions. Table 6.1 lists a compilation of all of them so that when you encounter any of this terminology you know what neutron energies they refer to.

TABLE 6.1: Nomenclature of neutrons used by reactor physicists, nuclear engineers, nuclear technologists and nuclear physicists, based on kinetic energies of neutrons.

Kinetic energy	Temp $(T=\frac{K.E.}{k_B})$	Speed $\beta = v/c$	$\lambda = \frac{28.6\times10^{-6}}{\sqrt{E[MeV]}}$ [nm]	Nomenclature [neutrons]
$< 3 \times 10^{-7}$ eV	$\sim$ mK	$\sim 10^{-8}$	100	Ultra cold
$< 5 \times 10^{-5}$ eV	$\sim$ K	$\sim 10^{-7}$	10	Very cold
< 0.025 eV		$\sim 10^{-6}$	> 0.2	Cold
~ 0.025 eV	$\sim$ 300 K	$\sim 7 \times 10^{-6}$	0.2	Thermal
~ 0.2 eV	$\sim$ 3000 K	$\sim 10^{-5}$		Hot
~ 0.4 eV	$\sim$ 6000 K			Slow
< 1 eV	$\sim$ 11,000 K	$\sim 10^{-5}$	0.03	Low energy
1 eV to 10 keV	10^4 to 10^8 K	10^{-3}	$> 10^{-3}$	Epithermal resonance
< 1 MeV		0.04		Fast
10 keV to 25 MeV	10^8 K to 10^{12} K	0.2		Continuum

$$k_B = 8.617 \times 10^{-5} \text{ eV K}^{-1} \text{ or } 1 \text{ eV} = 11605 \text{ K}.$$

Please note that the energy ranges of each type of neutron are approximate. As you see, the lowest five types, i.e., ultra cold, very cold, cold, thermal and hot neutrons, derive their names from the concepts of thermodynamics.[3]

As order of magnitude estimates, one may remember
1 eV = 10^4 K, 1 keV = 10^7 K, etc.

Borrowing from this concept, neutrons of 0.025 eV kinetic energy are of the corresponding temperature

$$T = \frac{K.E.}{k_B} = \frac{0.025}{8.617 \times 10^{-5}} \sim 300 \text{ K},$$

roughly room temperature and are called thermal neutrons. They move at average speeds of about 20 cm/second. The cold, very cold and ultra cold neutrons refer to those with lower and lower energies. In modern times, neutron sources of sub-micro Kelvin temperatures are being researched. They would be almost at rest in the laboratory.

---

[3] See Section 6.6.

The terminology of slow and fast neutrons is used for those of a few eV to about a few keV. Clearly slow neutrons are faster than thermal neutrons.

The term resonance neutrons is more specific about the nature of physics interactions that neutrons of a few eV to a few keV energies are likely to undergo with materials. As we have seen above, resonances correspond to enhanced interaction, quite often by a factor of one thousand or more compared to interactions in neighboring energies. They effectively remove neutrons from the medium, though there is some resonant elastic scattering too.

The term continuum neutrons also refers to the physics of neutron interactions. Here also neutrons are likely absorbed but cross sections are much lower than those in the resonance region. This region corresponds to nuclear excitation energies where discrete level structures of nuclei are no longer apparent. If there are resonances, they are very broad and spread over a few MeV or so. The nuclear spectrum looks like a broad continuum and hence the name continuum neutrons leading to these levels.

Thus the nomenclature gives a rough idea of neutron energies being referred to and enables one to get a general feeling of the type(s) of interactions that neutrons of those energies may undergo in a medium. At all energies, neutrons undergo elastic collisions which vary monotonously with energy.

If a neutron lives long enough in a medium without being absorbed by any of the nuclear processes, its ultimate fate is beta decay by emitting a proton, electron and antineutrino. The characteristic half-life for such decay is 10.2 minutes. Thus neutrons are finally lost in the medium one way or another. Below we discuss a few physics phenomena relevant in each energy region.

6.1.2 Low Energy Neutrons

Let us consider neutrons of approximately thermal energies or lower. At these energies, they may still be absorbed by atomic nuclei. The other phenomenon is scattering by their interaction with atomic electrons. This phenomenon is heavily exploited by material scientists to study material properties and neutron imaging. Here one relies on the wave nature of neutrons.[4] For a particle, we can write the corresponding de Broglie wavelengths in terms of the temperatures as

$$\lambda_{\text{de Broglie}} = \frac{h}{p} = \frac{hc}{\sqrt{mc^2 k_B T}} \tag{6.1}$$

where we used the non-relativistic approximation of $p = \sqrt{2mE}$. Here, m is the mass of the particle and E is its kinetic energy.[5] Furthermore, we use the thermodynamical expression of $E = k_B T$ for kinetic energy.

For neutrons, with $m = 939.6 \frac{\text{MeV}}{c^2}$, we can write

$$\begin{aligned} \lambda &= \frac{hc}{\sqrt{2mc^2 E}} = \frac{1240 \, [\text{MeV} \cdot \text{F}]}{\sqrt{2 \times 939.6 \times E} \, [\text{MeV}]} = \frac{28.6 [\text{MeV} \cdot \text{F}]}{\sqrt{E} [\text{MeV}]} \\ &= \frac{28.6 [\text{F}]}{\sqrt{8.617 \times 10^{-11} T [\text{k}]}} = \frac{0.308}{\sqrt{T [\text{k}]}} [\text{nm}] \end{aligned} \tag{6.2}$$

Table 6.1 lists the de Broglie wavelengths of neutrons of different energies. From the discussion in Chapter 1, we realize that the wavelengths of probes should be matched to the dimensions of objects under investigation. From Table 6.1, we recognize that cold and thermal neutrons of about a nanometer wavelength are well suited for studies of crystal structures. To study diffraction off nuclei, we need gamma rays of a few MeV energy. Also, a few MeV neutrons will do the job, except that competing processes may complicate the task.

In addition, neutrons are readily absorbed by nuclei. The product nucleus, though it is formed just above the threshold for the re-emission of a neutron, it is likely to reach its own ground state by emitting a cascade of gamma rays, with total energy equal to a Q value of $[M(Z,A) + M_n - M(Z, A+1)]c^2$, where $M(Z,A)$ is the mass of the nucleus with which the neutron interacts, $M(Z,A+1)$ is the mass of the product nucleus and M_n is the neutron mass. All masses are expressed in energy units. Also, quite often, the product nucleus is radioactive.

---

[4]See Chapter 1 for "matter waves," where this topic was thoroughly discussed.
[5]See Section 6.6.

Example 6.2

A ^{59}Co nucleus absorbs thermal neutrons (0.025 eV) to produce a ^{60}Co nucleus.

$$M(^{59}\text{Co}) = 58.993195 \text{ amu}$$
$$M(^{60}\text{Co}) = 59.993817 \text{ amu}$$
$$M(n) = 1.008665 \text{ amu}$$

Thus Q = 0.008043 amu = 0.005049×931.5 = 7.49 MeV or equivalently, in terms of mass excess $\Delta(^{59}\text{Co})$ = −62.224, $\Delta(n)$ = 8.071 and $\Delta(^{60}\text{Co})$ = −61.644, with the same result, Q = 7.49 MeV.

Each neutron capture with ^{60}Co as the product results in a series of gamma rays, known as prompt gamma rays or capture gamma rays, with a total energy of 7.49 MeV. The ground state of ^{60}Co is unstable and its half-life is 5 years. ^{60}Co is the most commonly used isotope for radiation therapy, agricultural applications such as food irradiation and also in teaching and research.

Epithermal, resonance neutrons: These two names are used synonymously. They correspond to neutron energies where they are preferentially absorbed by narrow excitation structures in the product nucleus. This excitation in the product nucleus, referred to as a compound nucleus, is likely to emit neutrons, protons or other charged particles before it reaches a stable or a radioactive nuclear state. The resonance energies and absorption cross sections are specific to the nucleus under investigation. We cannot overemphasize that these structures differ from one isotope to the neighboring one for the same chemical element. While we can have a general idea that a heavier neutron-rich isotope does not readily absorb neutrons, the details can be determined only by experiment.

Fast/continuum neutrons: Again, these two terms refer to nearly the same group of neutrons. One derives its name from neutron speeds, while the other name reflects a common feature that these neutrons excite nuclei beyond their discrete resonance structures, i.e., nuclear continuum. In many ways, nuclear continuum is analogous to the excitations of atoms or molecules above their ionization potential. Atoms or molecules shed electrons at excitations above ionization potentials. Nuclei may emit protons, neutrons or other particles to yield a product nucleus different from the target nucleus. In these energy regions, particle emission of the compound nucleus is most likely, resulting in a product nucleus different from the interacting nucleus. It is also very likely that the product nucleus is radioactive. Quite often, more than one neutron is emitted by the product nucleus, resulting in proliferation of neutrons. This is a well known phenomenon is nuclear fission which is a source of neutrons and a lot of radioactivity.

6.2 Nuclear Fission

At this stage, we take a short detour to describe nuclear fission. To conceptualize the physics, it would be beneficial to look at the binding energy per nucleon plotted against mass number A (Figure 6.2). It is useful to recollect that binding energy per nucleon is the average energy required to liberate a proton or neutron from a nucleus. Nuclei with higher binding energies are more stable against the emission of a nucleon.

The conspicuous feature is that

$$\frac{B}{A} = \frac{[M(A,Z) - (Z \times M_H) - (N \times M_n)]}{A} \quad (6.3)$$

starts with zero for A = 1, gradually increases, with a few spike-like structures, at ^{4}He, ^{12}C, etc., reaches a maximum value of about 8.7 MeV near iron and decreases to 7.5 MeV for mass 238. Thus, very light nuclei such as ^{1}H and ^{2}H can fuse together to form heavier nuclei accompanied by energy release. This sequence can go up to ^{56}Fe with lighter nuclei fusing to heavier ones. At the other end, nuclei such as ^{238}U are of smaller binding energies or more energetic systems. They can disintegrate into medium mass nuclei of about A~100.

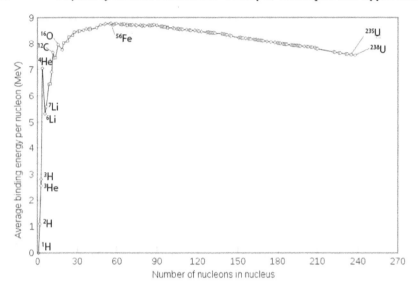

FIGURE 6.2: Plot of the binding energy per nucleon (B/A) against mass number. A few spikes for light nuclei (^{4}He, ^{16}O, etc.) and the broad maximum near A = 56 are noteworthy.

This process results in the energy release of about 200 MeV per fission. Both fission and fusion are examples where more energetic systems of smaller binding energies transform to less energetic systems of higher binding energies. The excess energies appear as kinetic energy, radiation, etc.

A simple picture may help visualize the process. This physical picture was due to Niels Bohr. Inside an atomic nucleus, protons and neutrons are in dynamic equilibrium. Heavy nuclei with an excess number of neutrons are not spherical but ellipsoidal in shape. A small distortion due to either internal motion or interaction with an external particle can disturb the dynamical equilibrium.

In nuclear fission a nucleus, driven to an unstable equilibrium, undergoes oscillations and splits into two or more massive nuclei along with emission of some lighter particles, mostly neutrons. Also, the fragments produced at the fission stage are neutron-rich and they undergo a series of beta decays to result in the end-product nuclei. Thus, unlike fusion, fission results in a lot of radioactivity. Some nuclei undergo decay by the fission process in the same way as nuclei undergo alpha decay, etc. This decay mode is known as spontaneous fission. Some isotopes of thorium, protactinium, uranium and transuranic elements exhibit spontaneous fission.

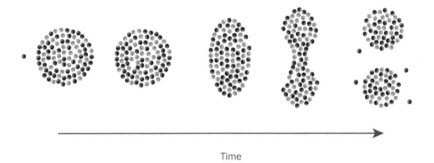

Time

FIGURE 6.3: Pictorial representation of neutron induced fission. A neutron approaches a neutron from the left. The compound nucleus is deformed and matter in it oscillates. The oscillations, when they are energetic enough, will rupture the nucleus into two smaller fragments.

Figure 6.3 is a pictorial depiction of fission dynamics. The uranium isotopes are deformed (they are not spherical in shape). A neutron incident on the ^{235}U target, if absorbed, forms a compound nuclear system of ^{236}U in an excited state. It tends to further deformation and matter is set in motion. If the motion is sufficiently violent, it can organize itself such that the compound nucleus becomes a system with two centers of mass with a narrow neck between them. As time progresses, the motion evolves such that the neck breaks, resulting in two nuclei and a few neutrons. The nuclei formed at the fission point are not stable nuclei. They undergo further beta decays to finally reach pairs of stable nuclei. The sequence for one specific decay chain is depicted in Figure 6.4.

In Figure 6.4, we see the reaction balance as

$$^{235}_{92}U + ^{1}_{0}n \rightarrow ^{140}_{54}Xe + ^{94}_{38}Sr + x \tag{6.4}$$

From the charge balance, we know that there is no charged particle missing. From the mass balance, we note that 2 amu are missing. Thus we conclude that two neutrons are simultaneously emitted in this

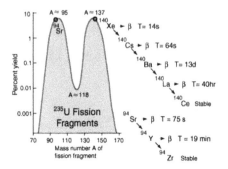

FIGURE 6.4: The relative frequency of the products resulting from uranium fission. (From HyperPhysics by Rod Nave, Georgia State University.)

fission decay. ^{140}Xe (Z = 54) undergoes a sequential decay before it ends in ^{140}Ce (Z = 58) and ^{94}Sr (Z = 38) ends in ^{94}Zr (Z = 40) in the following sequence.

$$^{140}\text{Xe} \rightarrow {}^{140}\text{Cs} + \beta^- + \bar{\nu}$$
$$^{140}\text{Cs} \rightarrow {}^{140}\text{Ba} + \beta^- + \bar{\nu}$$
$$^{140}\text{Ba} \rightarrow {}^{140}\text{La} + \beta^- + \bar{\nu}$$
$$^{140}\text{La} \rightarrow {}^{140}\text{Ce} + \beta^- + \bar{\nu}$$

and

$$^{94}\text{Sr} \rightarrow {}^{94}\text{y} + \beta^- + \bar{\nu}$$
$$^{94}\text{y} \rightarrow {}^{94}\text{Zr} + \beta^- + \bar{\nu}$$

A total of six beta particles and numerous gamma rays are involved in this process. In the sequential decay of mass 140, we note that ^{140}Ba has the longest half-life of 13 days, while its predecessors are of half-lives of a few seconds. Thus, the long term radioactive levels for mass 140 are determined by the activities of ^{140}Ba and ^{140}La. ^{140}La is in secular equilibrium with its parent ^{140}Ba. Of the two beta decays of mass 94, ^{94}Y is longer lived, with a half-life of 19 minutes. Thus the main source of long term radioactivity for this fission mode is ^{140}Ba and ^{140}La.

Example 6.3

We can calculate the energy release in the fission process by a standard method.

For the fission process

$$^{235}U + n \rightarrow {}^{140}Xe + {}^{94}Sr + 2n$$
$$^{94}Sr \rightarrow {}^{94}Y + \beta^- + \bar{\nu}$$
$$^{94}Y \rightarrow {}^{94}Zr + \beta^- + \bar{\nu}$$

The masses in atomic mass units are

$$M(^{235}U) = 235.04393$$
$$M(n) = 1.008665$$
$$M(^{94}Sr) = 93.915361$$
$$M(^{140}Xe) = 139.92164$$

$$Q = \left[M(^{235}U) + M(n) - M(^{94}Sr) - M(^{140}Xe) - M(2n)\right] \times 931.5 \text{ MeV}$$
$$= 0.22351 \times 931.5 \text{ MeV}$$
$$= 208.2 \text{ MeV}$$

Scientists immediately realized that this enormous amount of energy release is several orders of magnitude larger than other sources of energy and its potential for peaceful and destructive purposes.

The energy release is mostly kinetic energy, carried mainly by neutrons. Some of the energy release may be in the form of gamma rays. The product nuclei ^{94}Sr and ^{140}Xe decay to the end-products ^{94}Zr and ^{140}Ce, respectively. As $M(^{94}Zr) = 93.906315$ amu and $M(^{140}Ce) = 139.90544$ amu,

$$Q\left(^{94}Sr \rightarrow {}^{94}Zr\right) = [93.91536 - 93.906315] \times 931.5 \text{ MeV}$$
$$= 8.426 \text{ MeV}$$
$$Q\left(^{140}Xe \rightarrow {}^{140}Ce\right) = [139.92164 - 139.90544] \times 931.5 \text{ MeV}$$
$$= 15.09 \text{ MeV}$$

This energy release is mainly carried away as gamma radiation and kinetic energies of beta particles and antineutrinos.

The fission process in which two heavy fragments are emitted is called binary fission. In ^{236}U binary fission, the combination of ^{94}Sr and ^{140}Xe is one, among many, possibilities. Figure 6.4b shows the frequency of mass distribution of pair fragments. From experiments, it is seen that about one in ten fissions results in a ^{94}Sr and ^{140}Xe pair.

Although binary fission in which two fragments are formed is common, tertiary and quaternary fissions with three and four fragments are also known. In the past four decades, the main source of the medical isotope ^{99m}Tc, a decay product of ^{99}Mo, has been the fission process. About 6% of ^{235}U fission results in the ^{99}Mo end-product.

Example 6.4

^{252}Cf yields about 3.8 neutrons per spontaneous fission (SF) decay. The half-life of ^{252}Cf is 2.65 years. Then

$$\lambda_{^{252}\text{Cf}} = \frac{2.196 \times 10^{-8}}{2.65} = 0.829 \times 10^{-8} \text{ s}^{-1}$$

Three percent of ^{252}Cf decays are by spontaneous fission. If the decay has several modes and we need to know the rate of emission for a specific mode, then the half-life of the specific mode is $t_{1/2}/x$ where x is the branching fraction, with total branching of all modes normalized to one.

The decay constant for spontaneous fission is

$$\lambda_{^{252}\text{Cf}} \times x = 0.829 \times 10^{-8} \times 0.03 = 0.025 \times 10^{-8}$$

One microgram of ^{252}Cf contains 0.024×10^{17} atoms. The number of fission decays is $2.4 \times 10^{15} \times 2.5 \times 10^{-10} = 6 \times 10^{5}$ decays per second per μg of ^{252}Cf.

The yield is 3.8 neutrons per fission, so in one second there are about 2.3 million neutrons emitted due to SF of ^{252}Cf per μg. This isotope is used as a good source of fast neutrons for medical and industrial applications.

6.2.1 Chain Reactions

In a typical neutron induced fission of a ^{235}U nucleus, there are two free neutrons in the final state. The free neutrons are capable of creating secondary fission events in the second step, releasing four neutrons to cause fission in the third step and so forth. Thus there are 2^{n-1} neutrons causing the same number of fissions in the nth step. Since the reaction is propagating more reactions, a "chain reaction" is said to occur. If we let this continue, we have a runaway situation. We cannot afford it. Also, if we aggressively remove neutrons, we may not have sustained reactions and the process will die down. We cannot afford this either if we intend to use this process for applications.

TABLE 6.2: Some nuclei decay by spontaneous fission (SF).

Isotope	Half-life	SF branching
^{238}U	4.5×10^9 years	$\sim 1 \times 10^{-6}$
^{256}Fm	3 hours	0.9
^{256}Cf	12.3 minutes	~ 1
^{252}Cf	2.65 years	0.03

A major task of nuclear reactor design is to control the number of neutrons and the number of fissions to avoid catastrophe, while ensuring that there is a stable fission condition. A reactor is said to be "critical" if, for each fission, there is one neutron causing a subsequent fission. If there is more than one neutron causing fission, the reactor is "supercritical" and if it is less than one, the reactor is "subcritical."

While we send in a thermal neutron, the product neutrons are not usually thermal. Indeed, they are very fast neutrons. The fission cross section for fast neutrons on ^{235}U is not large.

6.2.2 Neutrons in a Nuclear Reactor

A nuclear reactor, either a power reactor or a research reactor, is designed to operate in a controlled manner. This would entail designing the system such that neutron interactions at each point in the reactor are known and controllable. The task is to organize the geometry and materials in the reactor container to be able to predict the number of neutrons and their energy distributions at any time and location. While the design itself involves very extensive numerical simulations,[6] the basic principles are simple. We require enough slow or thermal neutrons to ensure a sustained fission. As the fission neutrons are generally fast ones, we have to slow them down. In their path to low energies, some of them would be absorbed. We should be able to estimate this probability and ensure that there are enough neutrons coming to low energies to cause fission. This is known as neutron transport in the medium.

In nuclear reactors, one is generally concerned about the slowing down processes of neutrons as well as their absorption, as one needs to design systems of chain reactions which can be controlled in a predictable way.

Moderators are required to slow neutrons efficiently to thermal energies without absorbing them, while neutron absorption materials would be needed to stop the chain reaction process at the will of an operator.

The slowing down process is accomplished by the elastic scattering of neutrons with atoms in the medium. For elastic scatterings, energy loss at a single encounter varies with the mass number of the medium. It can be derived from simple Newtonian mechanics of elastic collisions. The maximum transfer of energy, corresponding to the minimum energy of outgoing neutrons (E'_{min}), occurs during head-on collisions. The ratio of E'_{min} to the kinetic energy (E) of incident neutrons can be written[7] as

$$\frac{E'_{min}}{E} = \left(\frac{A-1}{A+1}\right)^2 = \alpha \qquad (6.5)$$

$$E'_{min} = \alpha E \qquad (6.6)$$

where A is the mass number of the scattering nucleus.

The energy loss of a neutron varies from zero to $(1-\alpha)E$. Zero energy loss corresponds to the situation where a neutron, after scattering, proceeds along its initial direction of motion (zero deflection). The lowest energy corresponds to a scattering angle of 180 degrees.

---

[6]To the best of my knowledge, Monte Carlo numerical simulation techniques were first developed for nuclear reactor design studies.

[7]See Section 6.6.

If we assume that the scattering probability is the same for all angles of scattering (isotropic distribution of scattered particles), then the average energy loss $(< \Delta E >)$ is simply given by

$$\langle \Delta E \rangle = \frac{(1 - \alpha)}{2} E = \frac{2AE}{(A + 1)^2} \tag{6.7}$$

The average energy of a neutron after a single collision is

$$E - \langle \Delta E \rangle = E - \frac{2AE}{(A + 1)^2} = E \left(\frac{A^2 + 1}{(A + 1)^2} \right) \tag{6.8}$$

A few things to note here: As is to be expected, head-on collisions result in maximum energy loss. Light nuclei are most effective in slowing down neutrons. For a hydrogen medium, a single head-on collision may bring a neutron nearly to rest. The average kinetic energy of a neutron after a single collision in hydrogen is half of its initial kinetic energy. For heavy nuclei, slowing down is a many collision process.

For example, with ^{238}U as the stopping material, $< E > = 0.98E$ or a single encounter will result in about 2% energy loss.

The average kinetic energy after n collisions is

$$E_n = E \left(\frac{A^2 + 1}{(A + 1)^2} \right)^n \tag{6.9}$$

We can estimate n, the average number of collisions a neutron encounters before the average energy is reduced to E_n, from an initial value of E for the neutron moving in a medium of mass number A.

$$n = \frac{\log \left(\frac{E_n}{E} \right)}{\log \left(\frac{A^2 + 1}{(A + 1)^2} \right)} \tag{6.10}$$

To understand the significance of these equations, we may consider fast neutrons of about 1 MeV energy slowing down in a nuclear reactor to thermal energies of 0.025 eV. There are moderators in the medium as well as nuclear fuel, which is uranium. A 1 MeV neutron, on the average, undergoes 25, 30 and 2090 collisions in hydrogen, deuterium and uranium materials before it is thermalized. This statement assumes that a neutron undergoes purely elastic collisions. At first look, heavy water (D_2O) may not look too advantageous. However, the situation is not so straightforward, as we have not yet considered neutron induced nuclear reactions resulting in absorption and thus loss of neutrons as they propagate in media. We will consider this aspect below.

Before we leave this discussion, we should introduce an interesting term of the neutron physics community, "neutron lethargy." In a specific medium, A is constant. We may then write Equation 6.9 for kinetic energy after n collisions as

$$E_n = E_0 k^n \tag{6.11}$$

where $k = \frac{(A^2+1)}{(A+1)^2}$ and E_0 is the kinetic energy of incident neutrons before the first collision.

We then find that the energy after successive collisions, say $n-1$ and n, is simply

$$
\begin{aligned}
E_n &= E_0 k^n \\
E_{n-1} &= E_0 k^{n-1}
\end{aligned}
$$

or the ratio of energies of neutrons after two successive collisions

$$\frac{E_n}{E_{n-1}} = k \tag{6.12}$$

independent of the energy of incident neutrons.

We can also write

$$\frac{E_n}{E_0} = k^n \tag{6.13}$$

$$\log\left[\frac{E_n}{E_0}\right] = n \log[k] \tag{6.14}$$

Reactor physicists introduce a new variable called lethargy (u) defined as

$$u = -n \log(k) \tag{6.15}$$

or

$$k^n = e^{-u} \tag{6.16}$$

This suggests

$$E_n = E_0 e^{n \log(k)} = E_0 e^{-u} \tag{6.17}$$

an exponential decrease of kinetic energy.

The ratio of the average energies after two consecutive collisions is the constant k. The fractional energy loss of a neutron during a collision does not depend on its previous history.

In reactor physics, one is interested in knowing the number of collisions a neutron of specific energy undergoes for slowing down to thermal energies. We may write, in terms of lethargy,

$$E_{\text{thermal}} = E_0 e^{-u} \tag{6.18}$$

We may also write

$$\log\left(\frac{E_n}{E_{n-1}}\right) = \log E_n - \log E_{n-1} = \log k = \text{constant} \tag{6.19}$$

Thus the logarithmic energy decrement between two subsequent collisions is independent of energy.

Neutrons lose most of their energy in the first few collisions and then they linger. A neutron becomes more lethargic as it loses energy, an apt term as neutrons move slower and slower with decreasing energy. For example, if we start with 1 MeV neutrons and $k = 0.5$ (hydrogen medium), then the energies of neutrons become $1, 0.5, 0.25, 0.125,$ $0.0625, 0.03125, \ldots$ MeV. The amount of energy loss in a collision is half of that in the previous collision.

6.3 Macroscopic Cross Section

So far, we have been discussing interactions at nuclear or atomic levels and interactions of neutrons with individual entities. In the real world we deal with compounds or mixtures. As in neutron environments, we deal with bulk matter.

The concept of macroscopic cross section (L^{-1} dimensions) comes in handy. In many cases, we want to know how thick a shielding medium or a moderating column should be as we design reactor assemblies, experimental arrangements or shielding for health and safety of public and personnel. We may write the intensity of a neutron beam as it traverses a distance, x, in the medium as

$$I(x) = I_0 e^{-N\sigma_t x} \tag{6.20}$$

where

$$\sigma_t = \sigma_{el} + \sigma_{inel} + \sigma_{\text{reaction}} + \cdots \tag{6.21}$$

is the total cross section of the interaction of neutrons in that medium via elastic, inelastic, reaction, etc. Here N is the number of scattering centers (atoms) per unit volume.

One defines the macroscopic cross section as

$$\Sigma = N\sigma_t \tag{6.22}$$

It has dimensions of inverse length and $1/\Sigma$ is the mean free path, a measure of the average distance over which the neutron flux drops to $1/e$ (36.8%) of the initial value. It plays the same role as radiation length for electrons and high energy photons. The mean free path of neutrons can strongly vary with a change in neutron energies.

For a mixture of elements

$$\Sigma = \sum_i \Sigma_i \tag{6.23}$$

In a compound, the atom density of each element is

$$N_i = \frac{\rho N_0 n_i}{M} \tag{6.24}$$

where n_i is the molar fraction of the ith element.

The macroscopic cross section is

$$\Sigma = \frac{\rho N_0}{M} \left(\sum_i n_i \sigma_i \right) \tag{6.25}$$

Here σ_i is the neutron cross section of the ith component.

If we express the density (ρ) in g/cm^3, use Avogadro's number,

N_0, (6.022×10^{23}), M the molecular weight, and σ the cross section in barns (1 barn $= 10^{-24}$ cm^2) we will have

$$\frac{\rho N_0 \sigma}{M} = \frac{\rho[g/cm^3] \times 6.022 \times 10^{23} \times \sigma \, [barns]}{M} \tag{6.26}$$

$$= \frac{\rho \times 0.6022\sigma}{M} \, [cm^{-1}] \tag{6.27}$$

or

$$\Sigma = \rho \times 0.6022 \sum_i (n_i \sigma_i) \tag{6.28}$$

Example 6.5

In natural UO$_2$ ($\rho = 10$ g/cm^3 and $M = 270$) 99.3% of natural uranium is ^{238}U and 0.7% is ^{235}U. Thus, $n_i = 0.007$, 0.993 and 2.0 for ^{235}U, ^{238}U and oxygen, respectively, for one UO$_2$ molecule. For 1 MeV neutrons, the cross sections are 6.84, 7.10 and 8.22 barns, respectively, for ^{235}U, ^{238}U and O$_2$ (from nuclear reaction data). Thus

$$\Sigma = 10 \times 0.6022 \, [0.007 \times 6.84 + 0.993 \times 7.10 + 2.0 \times 8.22]$$
$$= 0.525 \text{ cm}^{-1}$$

The mean free path is the reciprocal of the macroscopic cross section. Thus, in UO$_2$, the mean free path = 1.91 cm for 1 MeV neutrons. If the mean free path is much longer than the sample size, it is likely that most neutrons will escape the medium without interacting.

We should be careful in interpreting this result for the slowing down of neutrons. Nuclear cross sections can change significantly as neutron energies change. So one has to take those varying cross sections into consideration to be able to estimate the stopping distances of the propagation of neutrons reliably. Quite often, one resorts to extensive Monte Carlo simulations to this end.

To compare different media for the slowing down of neutrons, one defines moderating power as

$$M = \xi \Sigma_{\text{scat}} \tag{6.29}$$

where ξ^8 is equal to the logarithmic decrease of energy in the scatterer [$\log k$] and Σ_{scat} is the macroscopic scattering cross section.

Moderating power alone is not the criterion to determine if a medium is good for use as slowing down material. If a medium has a large absorption cross section ($\Sigma_{absorption}$), it will then be a neutron poison, where neutrons will be lost by nuclear reactions. Thus, a large Σ_{scat} is a necessary but not sufficient criterion for a good moderator. One has to look for material with a large value for the ratio of $\xi\Sigma_{scat}/\Sigma_{absorption}$, the ratio of moderating power and absorption probability. The larger this value is, the higher the probability is that a neutron slows down to thermal energies without being absorbed in the medium.

TABLE 6.3: A comparison of moderating power and absorption cross sections of some materials.

Medium	$\xi\Sigma_{scat}$	$\Sigma_{absorption}$	$\frac{\xi\Sigma_{scat}}{\Sigma_{absorption}}$
water	1.28	0.022	58
heavy water	0.18	8.6×10^{-6}	21,000
beryllium	0.16	1.23×10^{-3}	130

Table 6.3 lists moderating power and absorption cross sections of water, heavy water and berylliam media, three commonly used reactor materials. The advantage of heavy water as a moderator is clear. Though normal water has higher moderating power, the fact that its absorption cross section is more than a factor of one thousand larger than that of heavy water makes normal water less attractive as a moderator.

6.4 Neutron Multiplication in Bulk Matter

In a nuclear reactor environment, one is interested in determining the number of neutrons at any given point in the medium from the known data at a specific point, say, the center of a nuclear reactor core. Reactor physicists devised a four factor formula to this end. In this section,

---

[8]We show in the appendix that $\xi \approx \frac{2}{A+2/3}$, where A is the atomic/molecular weight of the constituent medium.

we introduce this formula and discuss its significance. A neutron interacting in a medium may result in more than one neutron, and neutron multiplication occurs. Nuclear fission is one such phenomenon. Others are (n, xn) reactions, where x can be 2 or larger. The most common are $(n, 2n)$ processes or neutrons may be absorbed.

The four factor formula is a compact equation which captures the entire physics of neutron transport in four parameters. The multiplication factor k is defined as the ratio of neutrons in the $(i+1)$th generation to those in the ith generation. In reactors, the energies of neutrons range between 0.0001 and 10 MeV. At these energies, there are several possible interactions. Scattering is only one among them. With ^{238}U as the medium, fission can occur for neutron energies $E_n > 1$ MeV. Below those energies, neutron absorption occurs. One then defines a factor

$$\eta(E) = \frac{\text{\# of fission neutrons produced}}{\text{\# of neutrons absorbed}} = \frac{\nu \Sigma_f(E)}{\Sigma_a(E)} \tag{6.30}$$

$\Sigma_a(E)$ is the sum of cross sections of all processes where neutron absorption occurs, $\Sigma_f(E)$ is the fission probability and ν is the number of neutrons produced per fission.

6.4.1 The Four Factor Formula

$$k_\infty = \varepsilon p f \eta_T \tag{6.31}$$

where k_∞ is the neutron multiplication factor in a reactor of infinite size. ε is the fast neutron fission factor ($\varepsilon > 1$). It is the ratio of the number of neutrons produced by fission of all energies divided by the number of neutrons produced at thermal energies.

$$\varepsilon = \frac{\text{\# of neutrons from all fissions}}{\text{\# of neutrons from fissions due to thermal neutrons}} \tag{6.32}$$

p is the resonance escape probability ($p < 1$). At low and intermediate energies, neutron interactions show resonance structures. When a resonance capture occurs, the neutron is lost from the flux. p represents the fraction of neutrons which escape the resonances to reach thermal energies. A fraction $f (f < 1)$, known as the thermal utilization factor, enters the fuel and they are absorbed there. For each thermal neutron absorbed, $\eta_T (> 1)$ fission neutrons result. Thus, if there are n neutrons in the ith generation, there will be kn neutrons in the $(i+1)$th generation.

The four factor formula captures the entire physics of multiplication for an infinite sized reactor.

p specifies the probability that the neutrons so produced continue to lose their energies without being absorbed in the medium. f is the fraction of the neutrons which slow down to thermal energies and cause fission and η_T is the number of neutrons generated by thermal neutrons. The product of $\eta_T \varepsilon$ gives the total number of neutrons produced due to fission of neutrons of all energies.

If $k = 1$, the number of neutrons is unchanged from a generation to the next one. The system is said to be critical. If $k > 1$, the number of neutrons increases exponentially and the system is "supercritical." If $k < 1$, the number of neutrons decreases as the reaction progresses, and it will eventually die down. The system is then "subcritical." In reactor physics and the engineering community, one often refers to a reactivity coefficient $(k-1)$. The statements above hold good for the reactivity coefficients of zero, positive or negative.

In recent times, reactor design calculations have become more sophisticated and complex simulations are carried out. They make use of microscopic cross section data to model neutron transport. They rely heavily on the data that nuclear physicists measure in the laboratory. As we have seen several times in this book, the National Nuclear Data Center and its counterparts in other parts of the world are a resource of vast amounts of data.

6.5   Questions

1. Neutrons of 1 MeV kinetic energy scatter off a target atom. Calculate the minimum energy of the scattered neutron if the target is

    (i) hydrogen

    (ii) heavy hydrogen (^{2}H)

    (iii) iron (A = 56)

    (iv) tin (A = 120)

2. For the above question, find the number of collisions required to reduce the kinetic energy of neutrons to 0.03 eV.

3. Commercial steel is made of 2% Mn (A = 55), 18% Cr (A = 52), 10% Ni (A = 58) and 70% Fe (A = 56) by weight, ignoring the small amount of carbon.

   The total neutron cross sections of 1 MeV neutrons are 1, 10, 8 and 3 barns for Mn, Cr, Ni and Fe, respectively.

   (i) Calculate the mean free path of 1 MeV neutrons in steel. It is desired to attenuate the neutron flux to 50% of the incident beam.

   (ii) Calculate the thickness of steel (in cm) to be used to this end. Assume the effective molecular weight of steel is M = 56.

4. The ^{239}Pu medium has the following cross sections for neutrons of 2 MeV kinetic energy:

   Elastic scattering cross section: 3.45 barns

   Absorption cross section (fission and other reactions together): 2.04 barns

   (i) Calculate the macroscopic cross section and thus deduce the mean free path of 2 MeV neutrons in this medium.
   The density of ^{239}Pu is 19.74 g cm^{-3}.

   (ii) If the absorption cross section were not contributing to the processes, how many collisions will neutrons undergo to slow down from 2 MeV to thermal energies (0.03 eV)?

6.6 Endnotes

Footnote 1

As neutrons are of zero electric charge, one would have expected that they do not interact with electrons at all. However, they exhibit magnetic interactions with electric charges. It is experimentally known that neutrons possess a magnetic dipole moment of -1.92 nuclear magnetons (a nuclear magneton $\mu_N = \frac{eh}{m_p c}$ is about 0.05% μ_B, the Bohr magneton, unit of electron magnetic moment). Present-day quark models which treat neutrons and protons as composite bodies made up of quarks and gluons offer a conceptual basis for the finite neutron magnetic moment. We can consider neutrons as consisting of an equal amount of positive and negative electric charges (one up quark of $+2/3$ e and two down quarks each of $-1/3$ e electric charges) with zero net charge. This is somewhat similar to the picture of a neutral atom. If charge distribution within a neutron is not uniform, there may be a net electromagnetic interaction and it can manifest as due to circulating charges or a spin effect. You may then wonder how a photon, which is also electrically neutral and of zero magnetic moment, interacts with anything at all. Theory is mum except to say that a photon is an electromagnetic quantum.

Footnote 3

Thermodynamics prescribes that molecules moving freely in a gas container have a distribution of energies. The temperature of a container is related to the average kinetic energy of molecules inside it by the relation

$$\text{Average kinetic energy} = 3/2 \, k_B T,$$

for all species of molecules independent of their chemical nature.

In statistical physics treatments, the factor 3/2 is dropped as we adapt the Boltzmann distribution. The Boltzmann constant, k_B, is

$$k_B = 1.38 \times 10^{23} \text{ J K}^{-1}$$
$$= 8.617 \times 10^{-5} \text{ eV K}^{-1}$$
$$= 0.086 \text{ meV K}^{-1}$$

For a particle of mass m, in Newtonian kinematics, we may write

$$v = \sqrt{\frac{1 \times K \cdot E}{m}} \tag{6.33}$$

Footnote 5

It turns out that Newtonian approximation is valid for these energies. In fact, it is a good approximation for up to several tens of MeV of kinetic energies. Figure 6.5 shows β (v/c) vs kinetic energy in MeV units. Clearly, for up to about 100 MeV kinetic energies, the relativistic (Einsteinian) and non-relativistic plots are indistinguishable, the difference being less than about 7%. Then they deviate from each other and by about 500 MeV kinetic energy, the non-relativistic expression gives $\beta > 1$, an unphysical value. Though the non-relativistic expression is quite good for low energy, some precision measurements require that we employ the relativistic expression. When in doubt, use the relativistic equation.

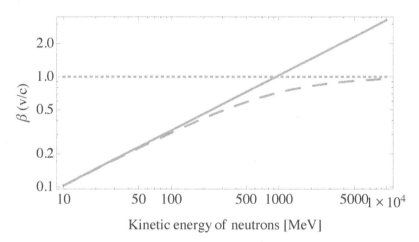

FIGURE 6.5: Plot of neutron speed β versus the kinetic energy of neutrons (MeV). The solid line is Newton's formula and the dashed line is the relativistic formula, which approaches the dotted limit of $\beta = 1$.

Footnote 7

The energy transfer equation can be derived as below. Let m be the mass of a neutron and m_A be the mass of a scattering nucleus, where A is its mass number. In a head-on collision, the neutron has the least energy when the scattering nucleus goes forward along the direction of the incident neutron and the neutron goes backward at 180 degrees. Labeling the initial and final speeds of the neutron as v and v_A as the speed of recoil atom A, we can write the momentum and kinetic energy conservation as

$$mv = Amv_A - mv'$$

$$\frac{1}{2}mv^2 = \frac{1}{2}m(v')^2 + \frac{1}{2}Amv_A^2$$

Solving these two equations to eliminate the velocity of unobserved recoil nucleus v_a, we get

$$v - v' = \frac{v + v'}{A}$$

or

$$\frac{v'}{v} = \frac{A-1}{A+1}$$

Thus the ratio

$$\frac{\text{Minimum energy of neutron}}{\text{Incident neutron energy}} = \frac{(v')^2}{v^2} = \frac{T'_{min}}{T} = \left(\frac{A-1}{A+1}\right)^2$$

7

Basics of Radiation Dosimetry

7.1 Introduction

While we mostly hear about radiation hazards to living beings, it can also affect inanimate objects, making transparent ones opaque, structures brittle and many other conspicuous changes such as color changes. It can induce permanent or temporary changes in the constitutions of these materials. In the modern space age, when satellite communications are essential to our way of living, a serious concern is radiation damage to satellite instrumentation by cosmic radiation. Van Allen radiation belts extending over 100–60,000 km above the earth's surface are known regions of high energy electrons and protons among others. The most recent finding shows that there are three radiation belts[1] causing excitement among the space and atmospheric science communities.

As we are concerned with radiation doses, we need to consider the overall effects of energy deposits, ionizations, chemical transformations and biological implications. Naturally, this topic is far beyond what physics can describe. However, physics can make quantitative statements about energy deposits and ionizations. In this chapter, we present these details and borrow the terminologies of radiation safety communities to become familiar with their practices.

So far, we discussed the behavior of a primary particle or quantum of radiation as it passes through a material medium. We presented that radiation, as it passes through a medium, deposits energy, ionizes the medium, etc.

There are two international organizations, the International Commission of Radiological Protection (ICRP, http://www.icrp.org/) and

---

[1] Visit http://science.nasa.gov/science-news/science-at-nasa/2013/28feb-thirdbelt

the International Commission on Radiation Units and Measurements (ICRU, http://www.icru.org/), which are concerned with the quantification of radiation effects on living systems and making recommendations on safe levels of radiation exposure. While there is no doubt that the organizations and their committees are committed to safety to make recommendations of universal validity, their work is very complicated. There are several contributing factors.

1. Radiation effects are both stochastic and non-stochastic nature.

2. For the same amount of energy deposits and ionizations, radiation damages vary with the type of radiation. For example, heavy charged particles cause more damage than photons or electrons for the same energy deposits.

3. Different organs respond differently to the same amount of energy deposits. Thus, radiation effects depend on which organ(s) receive doses.

4. When a source emits more than one species or one single energy radiations, the effects are cumulative. One has to work with average energies and/or total energies of emitted radiations.

5. If a radiation source is ingested by the body, the resulting radiation damage is estimated by taking into account the physical half-life of the radiation source and the biological half-life or the time it takes the body to get rid of the radiation.

While it may not be perfect, the international community has done a commendable job in providing good guidelines for safeguards.

There are a few terms specific to radiation dosimetry that we should be familiar with:

Exposure This refers to the ability of incident radiation to ionize a medium. To make a quantitative comparison, for reference standards, the quantity of ionization of an air medium by radiation is used. The unit of exposure is the roentgen. The amount of exposure is said to be one roentgen if the radiation liberates 2.58×10^{-4} C of electric charge in a kilogram of air.[2] This unit is no longer in much use, but the SI[3] unit of C/kg is currently recommended.

In a radiation field, one can estimate exposure by employing an ionization chamber and measuring the current flow through the chamber. One ampere of current passing through one kilogram or one milliampere of current through one gram of air medium amounts to a 1 C/kg exposure rate. Cumulative exposure can be measured by charge integrated over the time of exposure.

Dose This refers to the energy absorbed by the medium. The SI unit is a gray. The dose is said to be one gray (Gy) if the energy deposit is 1 J/kg. In non-SI (cgs) units, rad is the unit for dose. It was based on the definition of roentgen. As we see above,

$$1 \text{ R} = 2.58 \times 10^{-4} \text{ C/kg} \tag{7.1}$$

As electrons are of charge $q = 1.6 \times 10^{-19}$ C,

$$1 \text{ R} = \frac{2.58 \times 10^{-4}}{1.6 \times 10^{-19}} \frac{\text{charges}}{\text{kg}} = 1.61 \times 10^{15} \frac{\text{charges}}{\text{kg}}$$

If we know the ionization potential of air (energy required to free an electron), we can then relate the energy deposit measured in rads to the exposure measured in roentgens.

From the literature, the ionization potential of dry air = 33.97 eV.

We then have 1 roentgen resulting in an energy deposit of

$$
\begin{aligned}
1 \text{ rad} &= 1.61 \times 10^{15} \times 33.97 \text{ eV/kg} \\
&= 1.61 \times 10^{15} \times 33.97 \times 1.6 \times 10^{-19} \text{ J/kg} \\
&= 8.75 \times 10^{-3} \text{ J/kg}
\end{aligned}
$$

---

[2]See Section 7.5.

[3]Le Système international d'unités (SI) is the international system of units. A useful resource of the units, rules, styles, etc. is the website http://physics.nist.gov/cuu/Units

Thus, 1 rad is equivalent to a 0.0087 J/kg dose. One rounds off the result to 1 rad = 0.01 J/kg.[4]

This unit, then, is simply related to the cgs unit rad as 1 Gy = 100 rad.

Equivalent Dose As we deal with animate matter, we are interested in knowing not just energy deposits, but about the physiological effects that the radiation can produce. In some sense, this effect is a combination of energy transfer and ionization caused by the radiation, i.e., exposure and dose.

To this end, one defines an equivalent dose (H), which is the dose multiplied by a quality factor (Q), specific to the type of radiation and energy.

The unit is a sievert (Sv) = dose (Gy) $\times$ Q

A sievert is related to the cgs unit rem, 1 Sv = 100 rem.

The quality factor (Q) is defined as $Q \equiv 1$ for photons and electrons.

ICRP publication 92 specifies that *RBE*[5] *should be expressed in terms of the pertinent biological effectiveness of ordinary X-rays taken as 1 (average specific ionization of 100 ion pairs per millimeter of water or linear energy transfer of 3.5 keV per mm of water).*[6]

As we have seen in previous chapters, the energy loss of a particle, especially that of neutral radiations, does not result in the deposit of all the energy at the point of interaction. For example, a photon interacting by Compton scattering will have another photon leaving the scene, which may escape the medium without further interaction. A neutron captured in a medium may result in a shower of

---

[4]This rounding off deviates from the standard ionization potential estimate by more than 10%. One also estimates that the effective energy deposit in air may be as much as a 38–40 eV/ion pair when one considers recombination effects, etc. This should not cause too much concern if we use the conversions as order of magnitude estimates and stick to SI units.

[5]RBE: Radiation Biological Equivalent is a quality factor to account for the relative biological effect of a specific radiation.

[6]International Commission of Radiological Protection (ICRP) publication #92, published by Pergamon Press (2003).

photons, some of which may escape the medium. For dosimetric purposes, we need to know the energy deposits. To this end, we define another parameter, KERMA.

KERMA $\underline{K}$inetic $\underline{E}$nergy $\underline{R}$eleased per unit $\underline{MA}$ss of the material.

In dosimetry, this parameter is of the utmost importance. It is the energy released in the medium which causes damage to the system. There are two components. One is collision loss, which is a local deposit of energy to particles (mostly electrons) which will in turn lose energy in the vicinity of a primary particle trajectory. This is called collision KERMA (K_{col}). The second is radiative KERMA (K_{rad}).

KERMA is mostly relevant to photon beams and neutrons. Of course, relativistic electrons behave like photons for energy loss and thus KERMA is relevant for them too. To determine these parameters, we must become familiar with a few other terms.

Fluence The number of particles/quanta crossing a unit area.[7] If a particle beam has areal cross section A, and n is the number of particles traversing the medium, the fluence is

$$\Phi = \frac{n}{A} \tag{7.2}$$

The rate of fluence is flux (F)

$$F = \frac{\Phi}{t} \tag{7.3}$$

The energy fluence of a beam of particles is

$$\Phi_E = \frac{\sum_i n_i E_i}{A} \tag{7.4}$$

where n_i is the number of particles of energy E_i and the sum is carried over total particles.

The rate of energy fluence is energy flux Φ_E/t.

---

[7]Perhaps one is more familiar with the term "flux." Flux is the number of particles traversing a unit area per unit time. Fluence is the total number traversing per unit area. Fluence is flux integrated over the time of measurement or event.

Example 7.1

Thermal neutron capture of neutrons by ^{14}N in tissue.

When a neutron interacts with ^{14}N, the only reaction besides elastic scattering is the ^{14}N$(n,p)^{14}$C of $Q = 0.63$ MeV.

The cross section of this reaction is $\sigma = 1.84$ barns/atom.

This energy is shared between protons and the ^{14}C ion, in inverse proportion to the masses of particles in the final state. Thus, the proton carries about 93% of the kinetic energy (~ 0.59 MeV), while ^{14}C takes the remainder of the energy (0.04 MeV).

1 g of tissue[8] contains 0.03 g of ^{14}N or 1.3×10^{24} atoms/kg.

KERMA per unit thermal neutron fluence (K/Φ) is equal to

$$\frac{K}{\Phi} = \sigma \frac{N}{m} \langle \delta E \rangle$$

$$= 1.84 \times 10^{-28} \times 1.3 \times 10^{24} \times 0.63 \times 1.6 \times 10^{-13}$$

$$= 2.4 \times 10^{-17} \frac{J}{\text{kg m}^2 \text{ neutron}}$$

that is,

$$\frac{K}{\Phi} = 2.4 \times 10^{-17} \text{ Gy/m}^2/\text{neutron}$$

A SLOWPOKE research reactor[9] operates at a flux of 5×10^{15} n/m^2/s. Thus

$$K = \tfrac{K}{\Phi} \times \Phi = 2.4 \times 10^{-17} \times 5 \times 10^{15} = 0.12 \text{ Gy/s}$$

Most often, we need to estimate the dose that a source of radiation will impart in a target volume. It may be a medical diagnostic using X-rays, gamma rays or it may be a radiation therapy treatment. It may also be a situation where we need to estimate the radiation levels to prevent accidental exposure to high levels.

---

[8]Annals of the ICRP, report 89 "Basic Anatomical and Physiological Data for Use in Radiological Protection: Reference Values," Ed: J. Valentin, published by Pergamon Press (2003).

[9]SLOWPOKE is the acronym for <u>S</u>afe <u>LOW</u> <u>PO</u>wer <u>K</u>ritical <u>E</u>xperiment, a research reactor developed by Atomic Energy of Canada Limited. A few reactors are still in operation.

It may seem a bit confusing since the doses are expressed in grays and sieverts in terms of joules per kilogram, while particle energies are in microscopic units of electron volts and intensities are in becquerels. A simple conversion will make things easy.

We shall illustrate this conversion by a simple example. Let us begin with radioactivity expressed in becquerels [A (Bq)]. Say each decay results in the emission of n_i radiations of the ith type each of energy E_i expressed in MeV units.

Example 7.2

Ninety-nine percent of decays of ^{60}Co result in the emission of a beta particle of average energy 0.096 MeV and photons of 1.17 MeV and 1.33 MeV. Then we have $n_\beta = 1$ and $E_\beta = 0.096$ MeV. We have $n_\gamma = 2$, $E_{\gamma_1} = 1.17$ MeV and $E_{\gamma_2} = 1.33$ MeV. For dose calculations, we are concerned the energy deposits in the specific volume, which may be a portion of the energy of incident radiations. Let us denote that the radiation n_i deposits a fraction ϕ_i of its energy in the target volume.

Say the mass of the specific volume is m kilograms. Energy per decay per unit target mass = $\frac{\Sigma_i n_i E_i \phi_i}{m}$ MeV.

For dose calculations, we have to convert to joules. 1 MeV is 1.6×10^{-13} J.

Also, we can calculate the dose rate per second $(\dot{D})$ by knowing the source activity (A). Or

$$\dot{D} = \frac{A \Sigma_i n_i E_i \phi_i}{m} \times 1.6 \times 10^{-13} \text{ Gy/s}$$

The ICRP report gives the dose in units of rads per day per microcurie on a 1 gram target.

We note that 1 day = 86,400 seconds, 1 microcurie = 3.7×10^4 Bq,

1 Gy = 100 rad and 1 kg = 1000 g. So

$$\dot{D} = \frac{A\sum_i n_i E_i \phi_i}{m} \times 1.6 \times 10^{-13}$$
$$\times 86400 \times 3.7 \times 10^4 \times 100 \times 1000 \text{ rad/day}$$

$$\dot{D} = \frac{A\sum_i n_i E_i \phi_i}{m} \times 51.2 \text{ rad/day}$$

$$\dot{D} = \frac{A\sum_i n_i E_i \phi_i}{m} \times 0.512 \text{ Gy/day}$$

7.2   Biological Considerations

Up to this point, it is straightforward physics. Beyond this, one has to resort to considerations of radiobiology to estimate radiation effects on living systems.

For photons and electrons, the quality factor is one. Thus, for X-ray or gamma ray imaging or radiation therapy with photons and electron beams,

$$\dot{D} = \frac{A\sum_i n_i E_i \phi_i}{m} \times 51.2 \text{ rem/day} \qquad (7.5)$$

$$\dot{D} = \frac{A\sum_i n_i E_i \phi_i}{m} \times 0.512 \text{ Sv/day} \qquad (7.6)$$

In the conversions to grays and sieverts, the mass must be expressed in kilograms, while it is expressed in grams for rads and rems.

For radiations of ions or neutrons, we have to multiply each species of radiation for a corresponding quality factor (w_{r_i}). Thus, the final expression reads

$$\dot{D} = \frac{A\sum_i n_i E_i \phi_i w_{r_i}}{m} \times 51.2 \text{ rem/day} \qquad (7.7)$$

$$\dot{D} = \frac{A\sum_i n_i E_i \phi_i w_{r_i}}{m} \times 0.512 \text{ Sv/day} \qquad (7.8)$$

where $w_{r_i} = 1$ for electrons and photons.

Example 7.3

A target volume of 5 g is exposed to radiations from 10 mCi of ^{60}Co for 1 day. Assume that β particles deposit all their energy in the target volume, while (i) all the energy of photons is deposited and (ii) 50% of photon energies escape to the surroundings. What is the total dose received by the target volume?

$$A = 10 \text{ mCi} = 10,000 \, \mu\text{Ci}$$

particle	energy
β	0.096 MeV
γ	1.17 MeV
	1.33 MeV

$\phi_i = w_i = 1$ for all radiations.

Then the summation term is $0.096 + 1.17 + 1.33$ MeV = 2.596 MeV.

(i) The dose per day is

$$\dot{D} = \frac{10,000 \times 2.596}{0.005} = 5.192 \times 10^6 \text{ Sv/day}$$
$$= 5.192 \text{ MSv/day}$$

(ii) If the target is such that half of the energy of photons escapes the volume, while electrons deposit their full energy, $\phi_\gamma = 0.5$ and $\phi_\beta = 1$ for gamma rays and electrons, respectively. Then the summation term is $0.096 + (2.5 \times 0.5) = 1.346$ and the dose is

$$\dot{D} = \frac{10,000 \times 1.346}{0.005} = 2.7 \text{ MSv/day}$$

In radiation therapies in cancer treatment, integrated doses of 20–80 grays or equivalently 20–80 sieverts of gamma rays are administered. For medical diagnostic examinations, the amount of radiation is a few millisieverts and it is almost always less than 20 mSv.

For radiological protection, a committee of the ICRP recommends two types of quantities, viz., tissue or organ equivalent dose (H_T) and effective dose (E).

The equivalent dose is defined as dose (in grays) times a quality factor (w_R), which is specific to each type of radiation. Thus an organ will be affected differently depending on whether the same amount of the dose is due to photons or neutrons or other charged particles.

Let us say a tissue T receives a dose of $D_{T,R}$ grays due to radiation of type R. Then the equivalent dose received by T is

$$H_T = w_R D_{T,R} \qquad (7.9)$$

In an environment with different types of radiation, we are concerned about the absorbed dose averaged over a tissue or organ and weighted for each radiation quality factor.

For example, we may be concerned about the exposure of a tissue to alpha, beta and gamma radiation emitted by a radioactive source. The equivalent dose is the sum total of doses due to all radiations.

The equivalent dose to the tissue is represented by H_T and is given by

$$H_T = \sum_R w_R D_{T,R} \qquad (7.10)$$

Equation 7.10 suggests that we calculate the dose due to each species of radiation separately, multiply them by their respective dose and add them all up to estimate the total equivalent dose to a target, either a patient undergoing medical treatment or a living being exposed to hazardous radiations.

Table 7.1 lists the radiation weighting factor of commonly encountered radiations, as recommended by the Nuclear Regulatory Commission of the United States of America. We note that electrons and photons are taken as unity. Neutrons of different energies have different values, with a maximum weighting for those of 0.1 to 2 MeV. It is worth noting that alphas and heavy ions cause the most damage as they interact by continuous ionization of the medium they interact with.

The next thing we would like to know is the total dose received by a body. This is obtained by multiplying equivalent doses to individual organs by corresponding tissue weighting factors (w_T)[10]. Tissue

---

[10]For bladder, liver, thyroid and oesophagus, $w_T = 0.04$ in 2007 and 0.05 in 1991.

TABLE 7.1: Radiation weighting factors (w_R) for various species of radiation. (From Informationskreis KernEnergie, Berlin www.kernenergie.de.)

Type of Radiation		w_R
Photons		1
Electrons and muons		1
Neutrons:	< 10 keV	5
	10–100 keV	10
	0.1–2 MeV	20
	2–20 MeV	10
	> 20 MeV	5
Protons	> 2 MeV	5
Alphas, fission fragments and heavy ions		20

weighting factors take into account that some organs are more sensitive to radiation effects than others. It is defined as the relative contribution of a tissue or organ to the health detriment resulting from uniform irradiation of the body.

The effective dose is the sum of weighted equivalent doses in all tissues and organs of the body.

$$E = \sum_T w_T H_T \tag{7.11}$$

The latest recommendations are from ICRP report #103, adopted in 2007. For comparison, Table 7.2 lists the w_T factors for various organs in the human body from the committee recommendations of 1991 and 2007.

In Table 7.2, we notice major differences for the breast and gonads between the 1991 and 2007 recommendations. These recommendations are criticized as being subjective by a few practitioners.[11]

---

[11] See, for example, http://www.icrp.org/docs/David%20Brenner%20Effective%20Dose%20 %20Flawed%20Concept.pdf

TABLE 7.2: The ICRP committee recommendations of tissue weighting factors (w_T) for a few organs. The recommendations of 1991 and 2007 are presented for comparison.

Organ	w_T (2007)	w_T (1991)
Gonads	0.08	0.20
Bone marrow	0.12	0.12
Breast	0.12	0.05
Colon	0.12	0.12
Lung	0.12	0.12
Stomach	0.12	0.12
Bladder, thyroid, liver, oesophagus	0.16	0.2
Bone and skin (0.01 each)	0.02	0.02
Brain and salivary glands	0.02	0
Remainder	0.1	0.05

Example 7.4

In ^{60}Co radiation therapy, a patient is not exposed to beta particles. Only photons enter the body. In each decay the ^{60}Co source emits two photons, one of 1.17 MeV and the second one of 1.33 MeV. Thus the average energy of a ^{60}Co photon is 1.25 MeV $= 2 \times 10^{-13}$ J. In a treatment, 10 billion photons deposit energy in a patient's liver weighing 100 g.

The equivalent dose is

$$H_T = \frac{\text{number of photons} \times \text{energy deposited per photon}}{\text{organ mass}}$$
$$= \frac{10^{10} \times 2 \times 10^{-13}}{0.1}$$
$$= 0.02 \text{ Sv}$$

From Table 7.2 we find that the tissue factor (w_T) is 0.04 for the liver. Thus the effective dose is

$$E = w_T H_T$$
$$= 0.04 \times 0.02 = 0.0008 \text{ Sv}$$
$$= 0.8 \text{ mSv}$$

for the liver treatment. One has to consider the dose to surrounding parts of the body since the radiation passes through the skin and intermediate body mass before it reaches the liver.

The total effective dose equivalent (TEDE) is the sum of external and internal exposures.

In North America, the protection agencies[12] stipulate maximum permissible doses for radiation workers as follows.

Body part	Sv/year
Whole body	0.05
Lens of the eye	0.15
Extremities	0.5
Skin	0.5
Embryo	0.005
Occupational exposure of a minor	0.005

For the general public, the maximum allowed dose is 1 mSv/year for the whole body.

Radiation from sources external to a target impart instantaneous doses, while those internal to the target either due to inhalation or ingestion of radioactive materials deposit energies over extended periods of time.

Two phenomena contribute to the decrease of radiation levels in the body. First is the physical half-life $(t_{1/2}^r)$ of the radioactive material with its characteristic decay constant (λ^r). In addition, biological systems eliminate foreign elements with a characteristic ejection constant (λ^b) and corresponding half-life $(t_{1/2}^b)$. The effective decay constant (λ^e) is simply the sum of those two phenomena.

---

[12]United States Nuclear Regulatory Commission (USNRC), Title 10, code of Federal Regulations, Part 20, Standards for Protection Against Radiation.

The effective decay constant is

$$\lambda^e = \lambda^r + \lambda^b \tag{7.12}$$

Or the effective mean life (τ^e) is given by

$$\tau^e = \frac{1}{\lambda^e} = \frac{1}{\lambda^r + \lambda^b}$$

$$= \frac{1}{\frac{1}{\tau^r} + \frac{1}{\tau^b}} = \frac{\tau^r \times \tau^b}{\tau^r + \tau^b}$$

It is worth noting that the effective mean life is shorter than biological and radioactive mean lives. It is understandable because the two processes, viz., excretion and radioactive decay, contribute to deplete the radioactive material from the body and thus they are more effective than either process by itself.

It is then possible to calculate the dose for each year following intake in either a medical procedure, a nuclear event or a radiation worker engaged in mines. As a general policy, the estimates are made for 50 years in the case of adults and 70 years for children from the time of exposure to radiations.

The United States Nuclear Regulatory Commission (USNRC) defines the terms[13]:

Committed Dose Equivalent (CDE) The CDE is the dose to some specific organ or tissue of reference that will be received from an intake of radioactive material by an individual during the 50-year period following the intake.

It is listed as CDE (H_T,50) to indicate that it is the equivalent dose in sieverts to the tissue T over a period of 50 years.

Committed Effective Dose Equivalent (CEDE)

The CEDE ($H_{E,50}$) is the sum of the products of the committed dose equivalents for each of the body organs or tissues that are irradiated

---

[13]See the USNRC regulations (ibid).

multiplied by the weighting factors (W_T) applicable to each of those organs or tissues.

$$H_{E,50} = \sum W_T H_{T,50} \tag{7.13}$$

The committed effective dose (D_c) is readily calculated if the effective dose mean life is known by integrating the dose rate since intake. The dose rate decreases with the decay constant λ_e. Thus the committed dose over a time period T_c is given by

$$D_c = \int_0^{T_c} D_0 e^{-\lambda_{e_t}} dt \tag{7.14}$$

$$= \int_0^{T_c} D_0 e^{-\frac{t}{\tau_e}} dt$$

$$= D_0 \times \tau_e \times \left[1 - e^{-\frac{T_c}{\tau_e}}\right] \tag{7.15}$$

The calculations are simplified if T_c and τ_e are expressed in one set of units (years, days or hours) in the exponential term and D_c, D_0 and τ^e are expressed in one set, even if different from the one used in the exponential.

For example, if we need to calculate the dose per hour after 50 years, express D_c, D_0 and τ^e in hours, while for the ratio in the exponential use years as the unit for both T_c and τ^e as $50/\tau^e$ (years).

The USNRC defines a committed effective dose as the total effective dose delivered over a lifetime, assumed to be 70 years for a child and 50 years for an adult after an intake. Thus, $T_c = 50$ and 70 years for adults and children, respectively.

For $\tau^e \ll T_c$, the exponential term is zero. We then have a simple result that the committed dose becomes

$$D_0 = D_0 \times \tau^e$$

The mathematical modeling is specific to each radioactive isotope as the dose strongly depends on the half-life of an isotope and the number and species of radiations and their intensities.

For each isotope, the committed effective dose equivalent (CEDE) is readily calculated as

$$\text{CEDE} = \text{activity} \times \text{effective dose coefficient}$$

where activity is the intake amount of radioactivity. The effective dose coefficients specific to each isotope are listed in ICRP databases. We use e(50) and e(70) for adults and children, respectively.

7.3   Working Level Month

In view of the fact that radiation workers either in mines or at some installations are exposed to radiation on a regular basis, concepts of "working level" and "working level month" are introduced.

For people working in uranium mines, the ingestion or inhalation of radon gas is of serious concern. Also, in many cities, the public is exposed to this radiation due to the natural background in buildings. In the decay of the long lived, most abundant ^{238}U chain, ^{222}Rn is an inert gas, easily escaping from the mineral and migrating to surroundings. It is of half-life $t_{1/2} = 3.8$ days and its decay products leading to the ^{206}Pb stable endpoint are radioactive.

The 4n+2 series diagram in Chapter 2 shows the decay series from ^{222}Rn to ^{206}Pb. In the decay process, four alpha particles and a few beta particles are emitted. We are not concerned with the beta particles in this case. ^{210}Pb decay has the longest half-life, 22.2 years. Thus, for radiation equilibrium calculations, we consider the decay to terminate at ^{210}Pb. Also, the branching of ^{218}Po to ^{218}At is 0.02%, with 99.98% proceeding via alpha decay to ^{214}Pb. This beta branching contribution to equilibrium can be neglected. The radon and its progeny emit three alpha particles with a total kinetic energy of 19.2 MeV.

The equilibrium factor of radon with its progeny depends on ventilation conditions, aerosol concentrations and various other factors which vary from one venue to another and from indoors to outdoors. After several attempts to define dangerous levels of radiation in terms of activities (picocuries per one liter of air), the National Council for Radiation Protection (NRPC) recommended focusing on the energy deposits of radiation as a criterion because for health considerations, exposure levels as expressed by energy deposits are the relevant parameters.

It is estimated that, on average, a radiation worker spends 170 hours per month at the workplace. The working level (WL) is thus defined as the radiation level releasing 130,000 MeV energy per 1 liter of air in 1 hour. In SI units, 130,000 MeV $= 20.8 \times 10^{-9}$ J.

Thus, 1 WL = 20.8 nJ/L of air.

In SI units, 1 WL = 20.8 mJ/m^3.

Thus Working Level Month (WLM) is defined as 1 WL exposure for 170 hours.

An earlier definition of working level specified WL to correspond to radon activity of 100 pCi per 1 liter of air.

100 pCi $= 100 \times 10^{-12} \times 3.7 \times 10^{10}$ Bq $= 3.7$ Bq.

The kinetic energy released by 3.7 Bq activity per hour is

$$3.7 \times 19.2 \text{ MeV} = 255,744 \text{ MeV/hour}$$

This is nearly two times the working level recommended by the NRPC.

If the progenies are in equilibrium with the ^{222}Rn parent, we would expect the dose levels to be almost a factor of two higher than the WL definition. The NRPC attributes this discrepancy to the fact that the progenies are not in secular equilibrium. Equilibrium correction factors F = 0.4 for indoor and F = 0.6 for outdoors are thus recommended. Understandably, we stick to the definition of WL with regard to the energy deposit as the dose has relevance. Thus 1 WL = 130,000 MeV/hour.

Workers are normally expected to spend a maximum of 170 hours in a month.

$$1 \text{ WLM} = 22.1 \times 10^6 \text{ MeV/month} = 22.1 \text{ TeV/month}$$

7.4 Specific Gamma Ray Constant

When a specimen is exposed to gamma radiations from a source, the dose depends on the following parameters:

- strength of the source
- geometry of the source

- media between the source and specimen at which the dose is calculated

- distance between the source and specimen

The energies and intensities of gamma rays emitted by radioactive nuclei are very specific to each radioactive isotope. Also, gamma ray attenuation varies with their energies. Thus, for a given source of radiation, the relative contribution of each gamma ray depends on the intensity (known as branching per decay of the nucleus). Once we fix the source of radiation, these parameters are fixed. Next we should consider the attenuation of photons in the medium which also vary with energies. Once we specify the medium and the distance in which the radiation travels, relative intensities and thus the relative contribution of each photon energy to the dose are also fixed.

Thus, once and for all, for each source of radiation and a specific medium, we can provide a specific gamma ray constant, also known as gamma factor, to be used. It was generally agreed to use air as the medium, perhaps due to ^{60}Co treatment or X-ray diagnostics one carries out. In the tradition of SI units, one meter is the reference distance. It was also agreed to use "hour" as the reference exposure time since the doses are always referred in rems/hr or Sv/hr. As for the distance, we find the constant referred to 1 m, 30 cm or 1 cm. Also, for the source activity, we find reference to MBq, mCi or Ci. Also, the doses are given for sieverts or rads.

We will calculate for a 1 MBq source at a distance of 1 meter and express the constant in sieverts. We then calculate the conversion to other units.

For activity expressed in mega Becquerels, we should multiply by 10^6 to get the dose calculation:

$$A[\text{Bq}] = AA[\text{MBq}] \times 10^6 \qquad (7.16)$$

To obtain dose/hour, we should multiply the activity contribution by 3600

$$t[\text{s}] = 3600\frac{\text{s}}{\text{h}} \qquad (7.17)$$

To get the dose in units of sieverts, we must express the photon energies in joules.

Energy (J) $= 1.6 \times 10^{-13} E_\gamma [\text{MeV}]$.

In a radioactive decay, the probability of the emission of a photon of energy E_γ is characteristic of the isotope. For almost all commonly known radioactive isotopes, these probabilities have been extensively measured and they are available on public domain websites.

They are generally expressed as branching ratios or intensity per 100 decays. For an isotope emitting several gamma rays, this factor is unique to each gamma ray energy and we can find it on those websites.

We should divide this factor, n_γ, by 100 to determine the contribution of a specific gamma ray to the dose.

As the gamma rays travel the distance between the source and the target, their intensities are attenuated by two factors.

One factor, common to all energies, is the decrease of intensity of radiation from a point source as the inverse square of distance (d).

$$I_d \propto \frac{1}{4\pi d^2} \tag{7.18}$$

This is simply the geometry factor of the inverse square law for intensity decreases in a homogeneous space.

The second factor, specific to each gamma ray energy, is attenuation in the medium due to photon interactions. We should note it does not depend on the isotope, but the characteristic of the gamma ray energy and the medium in which it travels.

For an arbitrary distance d, this reduction factor is

$$f^{medium}(E) = e^{\mu\left(E_\gamma \times d\right)} \tag{7.19}$$

$$D^{isotope}(E_\gamma) = A \times 3600 \times E_\gamma \times 1.6 \times 10^{-13} \times n_\gamma \times \frac{e^{-\mu E_\gamma \times d}}{4\pi d^2} \tag{7.20}$$

Collecting all the factors together, we can write the dose due to gamma ray emission of one specific energy from a radioactive isotope, in sieverts, as

$$D^{isotope}(E_\gamma) = 3.726 \times 10^{-7} \times E_\gamma[\text{MeV}] \times n_\gamma[\%] \times \frac{e^{-\mu E_\gamma \times d}}{4\pi d^2} \tag{7.21}$$

for the distance d in meters and an isotope sample of 1 MBq activity.

Note that the isotope property appears only in the energy of a photon and its intensity (number of gamma rays of that energy per 100

TABLE 7.3: Composition of the materials used for Figure 7.1.

Material	Carbon (Z = 6)	Nitrogen (Z = 7)	Oxygen (Z = 8)	Argon (Z = 18)
Composition	0.000124	0.755267	0.231781	0.012827

decays). The attenuation factor is characteristic of gamma ray energy. Once we specify the energy, the attenuation depends on the medium.

We can thus write the dose due to gamma radiation at a distance of 1 meter from a 1 MBq source as

$$D^{isotope} = 3.724 \times 10^{-7} \times \frac{\sum E_i \times n_i \times e^{-\mu_i \times d}}{d^2} \qquad (7.22)$$

where we sum the contributions to the dose from all gamma rays emitted by the source.

As mentioned above, the specific gamma ray constant is defined for an air medium. Below, we find an empirical formula for an air medium for energies of 1 keV to 24 MeV, which covers the range of photon energies commonly encountered in medical and health physics applications. The attenuation coefficients for arbitrary energies can be obtained from the XCOM website.

From Figure 7.1, we notice that the attenuation coefficient decreases rapidly with increasing energies for photons of less than about 30 keV and slowly decreases for higher energies. Therefore, we divide the energy region into two groups to make a fit of the data to empirical formulae. We find the best fit is obtained with

$$\mu(E_\gamma) = 2 \times 10^{-6} \times E_\gamma^{-2.788} \left[m^{-1} \right] \qquad (7.23)$$

For $E_\gamma < 0.03$ MeV

$$\mu(E_\gamma) = 0.0073 \times E_\gamma^{-0.437} \left[m^{-1} \right] \qquad (7.24)$$

For most energies, these fit formulae agree to better than 10% with the XCOM data.

The specific gamma ray constant for an isotope of 1 MBq is written as

$$\Gamma^{isotope} = 3.724 \times 10^{-7} \times \sum E_i \times n_i \times e^{-\mu_i} [\text{sieverts}] \qquad (7.25)$$

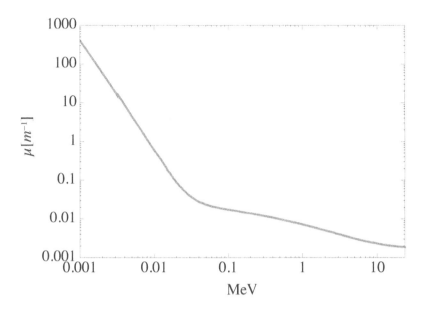

FIGURE 7.1: Attenuation coefficient (μ [m^{-1}]) for the air composition listed in Table 7.3. See the text for the empirical formula.

Example 7.5

A ^{60}Co isotope emits gamma rays of 1.17 and 1.33 MeV and each is nearly 100% branching. From the above, the attenuation coefficients for 1.17 and 1.33 MeV photons are

$$\mu(1.33) = 0.0073 \times (1.33)^{-0.437} m^{-1}$$
$$\mu(1.17) = 0.0073 \times (1.17)^{-0.437} m^{-1}$$

Thus

$$\Gamma^{60}{}^{Co} = 3.724 \times 10^{-7} \times 0.0073 \times \left[1.33^{-0.437} + 1.17^{-0.437}\right]$$
$$= 9.28 \times 10^{-5} \text{ Sv}$$

The above gamma factor is for 1 MBq activity.

We can easily convert to other conventions of specific gamma ray constants for 1 mCi (37 MBq) or 1 Ci (3.7×10^4 MBq) by multiplying by the corresponding factors.

$$\Gamma^{^{60}Co} = 1.378 \text{ Sv for a 1 Ci source}$$
$$= 1.378 \times 10^{-3} \text{ Sv for a 1 mCi source}$$

At times, we find the units rem used. Multiply the above numbers by a factor of 100 to this end.

Note: As we recognize, the conversion from curies to becquerels or sieverts to rems is very straightforward multiplication by a constant. In some cases, the specific gamma ray constant is referred to a distance of 30 cm or 1 cm. In those cases, we should pay attention to the d in the exponential term in addition to the inverse square dependence on distance.

Before we end this chapter, a few words of caution are warranted. The physics principles of radiation dosimetry are very straightforward, if they are not quite simple. But biological effects are a different matter. It is beyond my expertise to assess them. However, the changing standards of dosimetry and that there are commissions to establish guidelines are an indication that it is an ambiguous field of study. Also, in the literature and in practice, we find various notations, conventions and units. Thus, one has to exercise great care in arriving at quantitative estimates of radiation effects. Fortunately perhaps, we are mostly concerned with order of magnitude effects to which we apply further safety factors for our guidelines.

7.5 Endnotes

Footnote 2

The unit rad is derived as 1 roentgen = 1 electrostatic unit (esu) of charge per 1 cc of air.

$$1 \text{ esu} = 3.33 \times 10^{-10} \text{ Coulomb}$$

The density of air at STP is 1.29×10^{-6} kg/cc (1.29 mg/cc). Or 1 roentgen:

$$\frac{3.33 \times 10^{-10}}{1.29 \times 10^{-6}} \text{ C/kg} = 2.58 \times 10^{-4} \text{ C/kg}$$

8

Radiation Sources

8.1 Background

The first section of this book dealt with basic physics principles of nuclear radiation properties and their interactions with matter. Now we embark on some details of ways and means to produce, detect and quantify radiations. This is essential if we have goals for health, industry or scientific purposes or we need to assess the impact of species and quantities of radiation on the surroundings they interact with. The surroundings may be inanimate matter or live organisms.

Any arrangement involving nuclear radiations consists of some combination of the following components.

Radiation Sources: They may be natural sources such as cosmic radiations, background radiations in our surroundings, small radioactive sources, a particle accelerator or a nuclear reactor. In the case of cosmic radiations or backgrounds, our goal is generally to understand them and attempt to shield instruments and ourselves from them to minimize damage, if they are considered to be excessive. In modern times, we make radioactive sources or particle beams from accelerators with some specific scientific, technical or medical applications in view. Nuclear reactors for research or applications are also sources of radiations.

Material Medium: The radiation from a source passes through material media which either we insert in its path for some purpose or the medium happens to be there for reasons beyond our control. Regardless, radiations undergo several interactions, as we discussed in the first part of this book. This might itself be a source of secondary radiations, either serving a useful purpose or something we need to contend with. For example, an electron passing through a medium may emit bremsstrahlung radiation which, in turn, might create an electron-positron pair and so forth. An alpha particle moving in a material medium might yield a neutron due to a nuclear reaction.

Radiation Detectors: These are transducers which register the passage or absorption of radiation in them. The task of scientists and engineers is to optimize detector materials, geometries and configurations to achieve outputs to extract the desired information about radiations. The information we seek can be the number and species of radiation, their energies, times at which they pass etc. Depending on these applications, we design and build radiation detectors.

Electronics: The outputs from detectors are fed into electronic modules which are combinations of analog and digital signal processing units. They are configured to determine arrival times, energy deposits or spatial information of various radiations passing through the detectors.

Counting Systems: Counting systems are means by which we quantify the radiation information. In the time of Rutherford, he and his collaborators worked in dark rooms to count scintillations using the naked eye. In the assemblies of bubble chambers and cloud chambers, photographs were the counting systems. Today, electronic counters (also called scalers) and elaborate computer systems serve this purpose.

Also these days, as with everything else in our lives, computers play very significant and almost indispensable roles in controls and monitoring of radiation facilities as well as data analysis. The tasks range from design, construction and operation to that of interpreting final radiation interaction effects in research, medical or industrial operations.

A specific project may be as simple as a single Geiger counter measuring beta and gamma radiation background and alerting personnel by a chirping sound. It may be as complex as a large hadron collider[1]

---

[1] http://public.web.cern.ch/public/en/LHC/LHC-en.html

attempting to detect a Higgs boson in the ATLAS detector[2] as high energy particle beams collide against each other and produce a huge debris of radiations flying in all directions. The basic physics principles involved are still quite simple. It is the technical complexity which requires teams of talented scientists and engineers to build huge detectors and ancillary equipment and the programmed computer systems.

In the following chapters, we will be concerned with physics principles and techniques of radiation sources, detectors and electronics. We have already dealt with the physics of the interaction of radiation with matter, which will be useful as we consider detectors. The emphasis is on understanding the principles rather than a detailed construction of any single system.

In the early 20th century, Rutherford and his collaborators made use of alpha radiations from natural radioactivity as particle beams. Rutherford's experiments, leading to the discovery of the atomic nucleus, were conducted with such sources. Even earlier, perhaps unbeknownst to them, scientists were accelerating electrons when they built cathode ray tubes.

The main purpose of accelerators is to produce particle beams of higher energies. We cannot accelerate neutral particles, whether they are neutrons, neutrinos or other particles devoid of electric charge. Of course, we cannot accelerate photons. Higher energy photons simply are of higher frequencies moving at a constant speed c. A particle accelerator increases the speeds and kinetic energies of charged particles. At these facilities, we can produce higher energy neutral radiations or short-lived charged particles as secondary beams resulting from interactions of the primary charged particles with particles in a target material or colliding beam particles.

We may classify accelerators as

1. electrostatic (direct current, DC) machines or oscillating field (alternating current, AC) machines depending on the type of electric field we employ.

2. Linear accelerators or circular machines based on the geometry of arrangements. In circular machines, the particles are moving in spi-

---

[2]http://public.web.cern.ch/public/en/LHC/ATLAS-en.html

ral paths or closed circles while they move along straight line paths in linear machines.

Cathode ray tubes of electron beams, which became available at the end of 19th century, were the first examples of electrostatic linear accelerators. The cathode ray tube[3] itself has an extended history starting from about 1855, when Heinrich Geissler developed a mercury pump for vacuum tubes, to Sir J.J. Thomson's discovery of the electron in 1897. Cathode rays are electrons, with kinetic energies proportional to electric potentials causing fluorescence along these paths.

It takes very little energy to free electrons from atoms and crystal lattices, etc.[4] Currently, we know that electrons are readily liberated from metal surfaces either by heating (thermionic emission) or by applying electric fields (field emission) or by photoemission (photoelectric effect) among other means. Once the electrons are freed from the material, they are easily guided by the external potential, gaining speed and energy as they move toward the anode.

Almost immediately, scientists made X-ray beams as bremsstrahlung radiations of electron interactions in materials. It is of interest to realize that X-rays are secondary beams which are produced by the interaction of primary electrons, i.e., we first produce electron beams of specific energy and intensity. When electrons are made to fall on a material, X-rays will be produced.

The acceleration principle of a charged particle is the basic Lorentz force equation. The force F on a particle moving at velocity v of charge q and mass m due to an electric field E and/or magnetic field B is given as

$$\vec{F} = q\left[\vec{E} + \vec{v} \times \vec{B}\right] = m\vec{a} \tag{8.1}$$

where a is the acceleration.

It is the basic idea that a charged particle gains or loses energy as

---

[3]Old television sets and oscilloscopes were built with these tubes as their main components. More recently, plasma screens and liquid crystal displays (LCD) are replacing them.

[4]One defines "work function" as the energy required to free electrons from materials. You can think of it as a synonym for "threshold energy," which we referred to for nuclear processes. The work function is usually a few eV or less for most common materials. Indeed, solar cells function on the principle that visible light (2–3 eV) can liberate electrons to induce electric current.

it traverses an electric potential gradient. Positive charges gain energy when they are directed to negative potentials and negatively charged particles gain energy when they travel toward positive potential. Magnetic fields neither accelerate nor increase energies of charged particles.[5] They are essential components in guiding and focusing particle beams to the point of interest.

A particle of one electronic charge $(q = \pm e)$ gains kinetic energy $T = eV$ after it passes through a potential difference of V volts.[6]

8.2 Production of Electromagnetic Radiation

The traditional method was production of X-rays as bremsstrahlung radiation emitted as a relativistic charged particle interacts with matter. As we saw earlier, electrons readily emit radiation as they pass through high Z material media. There are a few points to consider:

1. Bremsstrahlung radiation is a continuous spectrum with the end point at the tube voltage (V). The intensity of the emitted radiation is approximately proportional to $I_\gamma \propto \frac{1}{E_\gamma}$ for $E_\gamma \leq V$. We note that these tubes output radiation of both higher and lower energies than what one needs for a given application.

   Recognizing that the photon attenuation coefficient (μ) of a medium depends on the photon energy and that μ decreases with increasing energies for low energy photons, one employs beam attenuators to selectively remove low energy photons. The attenuators thus employed are called "beam hardeners."

   Example: The attenuation coefficients in a lead medium are $\mu = 344$ cm^{-1} for 30 keV photons and $\mu = 27$ cm^{-1} for 140 keV photons. If an X-ray beam of equal intensity for the 30 keV and 140 keV photons is passed through a 0.1 mm thick lead sheet, the ratio

---

[5]The betatron is the only particle accelerator which relies on a magnetic field to accelerate charged particles. Even this machine makes use of the fact that an alternating magnetic field induces an electric field.

[6]See Section 8.12.

of emerging photons is

$$\frac{I(30\ \text{keV})}{I(140\ \text{keV})} = \frac{e^{-344}}{e^{-0.27}} = \frac{0.032}{0.76} = 0.042$$

The intensity of 30 keV photons after traversing through 0.1 mm (100 μm) thick lead material is 4.2% that of 140 keV photons. Thus the beam is nearly pure 140 keV photons, while it was a mixture of equal intensities of two energies.

2. Electrons interact by emitting radiation and they also cause ionization. Ionization results in heat deposits in the medium it passes through. The design and construction of the instrument should ensure that there is no excessive heating.

Example 8.1

A beam of electrons of 100 keV energy and 10 mA current passes through a lead material of one radiation length thickness. The total power of the beam is 1 kWatt. In passing through one radiation length thickness, the beam reduces to 1/e (37%) of the power, i.e., power is dissipated at the rate of about 630 watts. This is like heat dissipated by a very bright light bulb in a room. Air cooling, water cooling or some other means must be arranged.

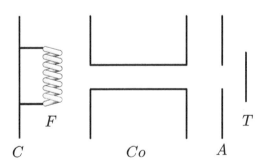

FIGURE 8.1: Idealized diagram of an X-ray production tube. C is the cathode, F is the electron source filament, Co is a collimator, A is the anode and T is the target from which the X-rays are emitted.

In a cathode ray tube, electrons are liberated from a hot filament, accelerated and then guided to the target material, which is the source of X-ray beams. An arrangement will similar to Figure 8.1.

The cathode (C) is kept at a negative potential with respect to the anode (A). F is a hot filament which emits electrons.[7]

The electrons produced in a hot element do not have a well defined direction. They may spread out in space. A combination of geometrical slits and collimators along with electric and magnetic fields are applied to render them into a parallel beam traveling along a well defined axis toward the anode.[8] The voltage difference (V) between the cathode and anode determines the kinetic energies of electrons to be equal to K.E. = eV. For example, if the potential difference between the cathode and anode is 100 kV, the kinetic energy of electrons is equal to 100 keV. Also, this determines the maximum energy of X-rays to be $E_\gamma^{max} = 100$ keV.

Example: Most commonly, we encounter X-ray machines in dental clinics. They are specified by the voltage difference between the cathode and anode in kilovolts, called the kilovolt peak (kVp), and electron current in milliamperes (mA), called the tube current.

A rating of 70 kVp means that the potential difference between the cathode and anode is 70,000 V and the maximum energy of X-rays is 70 keV. The current rating in milliamperes indicates the X-ray intensity. It should be noted that a higher current rating means more X-rays will be generated. If we keep the kVp at a constant value, the intensity of X-rays is proportional to the current. In general, if

---

[7]It turns out that electrons are easily emitted from many surfaces. They can be emitted by friction, by heating or by exposing them to light, etc., under certain favorable conditions. The criteria for choosing a material is that it should not be expensive, it should have a high melting point so that it does not melt upon heating and that it should be durable. In X- ray tubes, one generally employs a tungsten filament.

[8]The physics and mathematics of directing the particles are very similar to those of guiding and bending optical beams. For that reason, this area of study and research is called "beam optics." We define the focal length for particle beam systems just as we define the focal length of an optical system such as a camera lens or a telescope. In the same way as we use an arrangement of prisms to bend and direct optical rays, we employ magnetic fields to bend and direct charged particles. The terminology and mathematical apparatus are strikingly similar.

Inside a Dental X-ray Tube

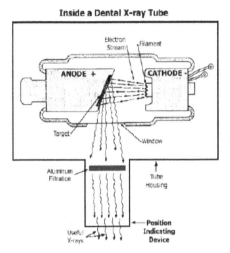

FIGURE 8.2: A schematic diagram of a dental X-ray machine. Electrons from the right fall on the anode. The anode geometry is such that most of the bremsstrahlung photons travel to the bottom port.

we increase the kVp, the X-ray intensity increases faster than the current.

Figure 8.2 is a schematic of a typical dental X-ray machine. In this scheme, the target is coated on the surface of the anode. Most of the energy is dissipated as heat and only a small fraction (about 1%) is converted into X-rays. As an example, a 100 kVp X-ray tube operating at 10 mA current outputs a power W,

$$W = 100 \, kVp \times 10 \, mA/ \sec = 100 \times 10^3 \times 10 \times 10^{-3} = 1000 \text{ watts.}$$

Clearly this is much hotter than household incandescent lamps. Thus, a cooling system surrounds the cathode-anode assembly to serve as a heat dissipater and insulator. Incidentally, it also absorbs some radiation. The arrangement is made such that X-rays exit to the outside of the tube through a narrow window. In the path of the X-rays an aluminum sheet is inserted. The recommended thickness of the aluminum sheet is 1.5 mm and 2.5 mm for voltage ratings below and above 70 kVp, respectively. This is explained below.

We know that the X-ray energy spectrum is a continuous bremsstrahlung with an end point energy at the kVp value. Only high energy X-rays, known as hard X-rays, penetrate a patient's tissue, and they are useful for imaging purposes. The low energy X-rays, known as soft X-rays, do not penetrate a patient's tissue. It is then useful to selectively deplete the soft X-rays within the instrument. The aluminum sheets serve this purpose.

Table 8.1 lists the mass attenuation coefficient (μ_ρ) in units of cm^2/g, the linear attenuation coefficient (μ) in units of cm^{-1} and the ratio of intensities after passing through 1.5 mm and 2.5 mm thick aluminum sheets.

The mass attenuation coefficients are taken from the XCOM database. The density of aluminum is 2.7 g/cm^3. The data is calculated for photon energies E = 10 − 100 keV in 10 keV steps.

From the table, we see that about 24 out of 1 million incident photons pass through the 1 mm Al filter, while intensities above 30 keV remain at 63% or higher levels. At 70 kVp operation, one works with X-rays of above about 50 keV energy. A 2.5 mm thick aluminum sheet reduces the intensity of photons below 20 keV to less than 10% levels, while maintaining the intensities of photons of above 50 keV at 78% or higher levels. This reduces the intensities of very soft X-rays (E < 10 keV) to below the 2×10^{-3}% level.

8.3 Ion Beam Accelerators

The first ion accelerators were built in the early 1930s. It is worth mentioning that George Gamow, a Russian èmigrè, constructed a theoretical model to describe the rates of alpha decay. As a consequence, it was deduced that one might be able to induce nuclear reactions if only one could produce ion beams of sufficient energies. The initial reaction of physicists was that it was an impossible dream. However, Cockroft and Walton combined their physics and engineering knowledge base to develop and build a particle accelerator and induce a nuclear reaction. In the last 80 years, the evolution of accelerator technologies to produce various species of primary and secondary particle beams is awe-inspiring. Not only do we produce beams of stable particles, we can now produce neutral beams and unstable charged particles with

TABLE 8.1: Coefficients of mass attenuation coefficient (μ_ρ) and linear attenuation (μ) for 10–100 keV X-ray in aluminum. The last two columns show the intensity ratios with and without 1.5 and 2.5 mm aluminum filters.

E (keV)	μ_ρ (cm^2/g)	μ (cm^{-1})	I/I_0 (1.5 mm filter)	I/I_0 (2.5 mm filter)
10	26.2	70.8	2.45×10^{-5}	2.07×10^{-8}
20	3.44	9.29	0.248	0.098
30	1.13	3.05	0.633	0.467
40	0.568	1.53	0.794	0.681
50	0.368	0.994	0.861	0.779
60	0.278	0.750	0.893	0.829
70	0.230	0.621	0.911	0.856
80	0.202	0.545	0.921	0.872
90	0.183	0.495	0.928	0.884
100	0.170	0.460	0.933	0.891

lifetimes as short as a few nanoseconds. For example, beams of pions (mean life, $\tau = 26$ nanoseconds) and muons (mean life, $\tau = 2.2$ microseconds) have been in use for science and applications.

A particle accelerator consists of three basic components:

(i) an ion source which produces charged particle beams

(ii) an accelerating field to increase energies

(iii) combinations of electric and magnetic fields to guide particles and focus to form particle beams of specific sizes

Right from the very beginning, both static electric fields and oscillating (alternating current) fields were candidates for accelerators and they were developed almost simultaneously.

8.4 Electrostatic Accelerators

Let us first discuss the electrostatic accelerator principle. It is based on the simple idea that a particle of electric charge q will gain an energy qV as it passes through an accelerating potential of V volts. As we want to increase energies, we have to increase the potentials proportionately. So, if we want to produce proton beams of 1 MeV, the potential should be 1 million volts and if we need 10 MeV protons, we need 10 million volts and so on. It is clear that we cannot keep on increasing electric potentials as the electrical insulation will break down.

Two early developments of electrostatic accelerators were (i) the Cockroft–Walton accelerator and (ii) the Van de Graaff accelerator.

(i) Cockroft–Walton Generator: This accelerator[9] is based on principles of voltage rectification and multiplication. It exploits the properties of diodes that they conduct electricity only when the cathode is negative with respect to the anode and that of capacitors[10] to store electric charge. The idea behind this system is to arrange a cascade network of capacitors and diodes connected to an alternating voltage. Thus it is also called a cascade generator or voltage multiplier. Each stage in this network is at a higher potential than the previous one. In an ideal system of n stages with a driving voltage of V volts, it can maintain $n \times V$ volts.[11]

Cockroft–Walton generators of high voltages of up to about 2.5 million volts have been built. Some modern accelerators, which comprise several acceleration stages, employ a Cockroft–Walton in the first stage to produce high intensity ion beams. Cockroft–Walton generators also find applications in several electrical and electronic gadgets such as laser systems, LCD backlighting and bug zappers, to name a few.

---

[9]John Cockroft (1897–1967) and Ernest Walton (1903–1995) were the co-inventors of this machine while they were at Cavendish laboratory, Cambridge, England.

[10]Electrical capacitors are commonly found in almost any electrical circuit. They are charge storage devices. An ideal capacitor, left to itself, can hold electric charge forever, but a practical capacitor holds a charge for finite times.

[11]See Section 8.12.

(ii) Van de Graaff accelerator[12]: Many of us might have come across hair-raising Van de Graaff potential generators either in classroom demonstrations or science museums. A Van de Graaff accelerator uses this principle to set up a high potential on a dome by charging it. The early versions consisted of an insulating belt moving between an electrical ground and an insulated dome which can be electrically charged. As the belt moves toward the dome, it comes in contact with a negatively charged electrical comb. The electrical comb strips the insulating belt of electrons, which renders the belt positively charged. The belt, in turn, loses the positive charge by capturing electrons at the dome, thus charging the dome. The belt returns to the negatively charged electrical comb and this process repeats. In theory, one may expect that this can lead to a dome which will have infinite charge and infinite potential. However, the electrical properties of the dome, the surrounding atmosphere, etc. limit it to a finite high voltage. Van de Graaff domes of voltages as high as about 15 million volts have been constructed.

In more recent versions of this accelerator, the charging rubber belt is replaced by a chain of conducting pellets separated by insulating materials. They are called pelletrons. But for that difference, Van de Graaff accelerators and pelletrons are identical in physics principle and operation.

As Van de Graaffs and pelletrons are still in much use, a brief statement about the tandem Van de Graaff or pelletron accelerators is useful.

We remind ourselves that it is the magnitude and relative signs (positive or negative) of ion charges and accelerating voltage which determine the acceleration of ions. Tandem accelerators use the same voltage (V) to accelerate charged particles to an energy of 2qV, where q is the magnitude of the electric charge of the ion. The principle is illustrated in Figure 8.3 and it is described here.

At point a the electric potential V = 0. We produce an ion of

---

[12]Robert Van de Graaff (1901–1967), an American physicist and professor at the Massachusetts Institute of Technology, was the inventor.

negative charge q. The negative ion has no kinetic energy at this point, i.e., E = 0. It sees the positive potential of voltage +V at point b and it is accelerated to point b. When it reaches point b, it has acquired kinetic energy E = qV. At that point, the ion is stripped of negative charges for it to aquire a positive charge +Q. The positive potential at point b is then a repulsive force and the zero potential at point c is attractive. When the ion reaches point c it would then acquire an additional energy of QV for a total energy gain of (q+Q)V in traveling from point a to point c.

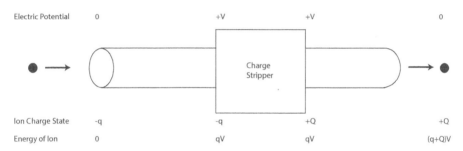

FIGURE 8.3: An ion of charge q is accelerated from point a held at potential V = 0 to b held potential +V. At point b, the ion is stripped of negative charge and a few more electrons to become a positive ion of charge +Q and it is accelerated to point c held at V = 0.

Example 8.2

A negative H^- ion of charge $q = -e$ is accelerated from point a to point b, the terminal of a Van de Graaff accelerator at a high voltage of 1 million volts. It would then gain an energy of 1 MeV as it travels from point a to point b. It is then stripped of two electrons to produce a H^+ ion (a proton). As it passes from point b to c, it will then acquire an additional 1 MeV kinetic energy, for a total of 2 MeV kinetic energy from a to c.

In the same arrangement, a singly charged negative ^{12}C ion ($^{12}C^-$) starts at point a. It acquires a kinetic energy of 1 MeV when it reaches point b. At point b, the ^{12}C ion is now stripped of all its electrons to create an ion of charge $q = +6e$. As it travels to point c from point b, it acquires an additional kinetic energy of $E = 6e \times V = 6$ MeV for a total gain of kinetic energy of 7 MeV as it traveled from point a to point c.

A main limiting factor of electrostatic accelerators to achieving higher energies is electrical breakdowns. A dry air atmosphere can hold potential differences of about 30 kV/cm or about 3 MV/m. There are other high insulation media. The most commonly used SF_6 (sulphur hexafluoride gas) is about five times as effective as dry nitrogen. However, this is still a finite, though definite, improvement in attaining higher energies.

8.5   Alternating Current Accelerators

Acceleration using oscillating fields poses interesting challenges compared to electrostatic fields. With electrostatic fields, all we do is set up an accelerating potential of the correct sign and the desired or maximum possible voltage. The particle will stream down the potential with increasing energy and we get our job done. With oscillating fields we should make sure that particles enter the field region to maximize the gain. However, not all particles have the same energy and thus they travel at different speeds, even though the differences are small. Oscillating fields are almost always a sine function of time. It may be worthwhile to examine field shape as a function at a point in the gap.

The field strength $E(t)$ at time t of a sinusoidal electric field with a maximum amplitude of E_0 is written as

$$E(t) = E_0 \sin{(2\pi f t)} \qquad (8.2)$$

where f is the frequency and E_0 is the maximum value of the oscillating field. In Canada and the USA, the household electricity is of frequency $f = 60$ Hz[13] and maximum voltages are $E_0 = 155 - 179$ V,

---

[13]1 Hertz (Hz) = one oscillation per second. Thus North American electrical supplies of $f = 60$ Hertz undergo 60 oscillations per second or 1 oscillation every 16.7 ms.

corresponding to the commonly discussed root mean square voltages of 110–127 V. In Europe and some other countries the maximum voltages are $E_0 = 311 - 339$ V, corresponding to 220–240 V_{rms} with a typical frequency of 50 Hz. For acceleration of particles, the voltages and frequencies are quite high. The frequencies, as we see below, are dictated by the particle type and they are in the radiofrequency ranges of a few megahertz to several gigahertz.

Based on the geometry of the arrangement, there are two types of machines.

Linear Accelerator In this arrangement, charged particles travel in straight lines. The electric fields are applied along the path such that the particles are in the field region when it is accelerating, positive for negatively charged particles and negative for positively charged particles. The alternating fields are of fixed frequency. At each passage a charged particle gains kinetic energy qV or its speed increases.[14] The design has to allow for this fact and ensure that a particle always sees an accelerating field (a positive field for a negatively charged particle such as electron or a negative field for a positively charged ion such as a proton, alpha, etc.). The particles should not see the oscillating field when it has a decelerating sign (field sign is the same as the sign of the charge of the particle).

Circular Accelerator Particles travel in spiral paths or closed single orbits. In this arrangement, particles enter the same accelerating field region once each revolution, allowing repeated use of the same accelerating power.

8.6 Linear Accelerators

For linear accelerators, the plan should be as follows. On average, particles arrive with an energy $E_{average}$ and they gain energy $\Delta E = E_0$. If we organize the fields and arrival times of particles of average energy to enter the field region at time t_0, when the field is E_0, then faster particles arrive early and they gain less energy, $\Delta E < E_0$. They get closer

---

[14]See Section 8.12.

to the energy of particles with average energy. While this is good, the
slower ones coming later will lag behind and they might even experience decelerating fields and be lost from the beam. If, on the other
hand, we arrange things such that particles with average energies arrive
when the field is E_c, slightly lower than E_0, then the faster ones come
when the field is $E_f < E_c$ and the slower ones come at a later time
$E_s > E_c$. Again the faster ones gain less energy, but now the slower
ones gain more energy than the average particles. Properly designed,
more particles will be of nearly the same energies and thus available
for further accelerations and/or manipulations by electric and magnetic
fields.

One usually optimizes the principal beam to arrive at a phase angle
$(2\pi f t)$ of 60 degrees, corresponding to $E_c \sim 0.86 \, E_0$, as a compro-
mise between maximum energy and the maximum number of particles
accelerated.[15]

Based on this principle, linear accelerators[16] for particles of various
species (protons, heavy ions, electrons, etc.) have been built. Currently,
many worldwide medical electron beam facilities of 4–15 MeV gamma
rays are linear accelerators. The longest electron linear accelerator is
located in Stanford, California. It is about 3 km long and it delivers
electron beams up to 50 GeV energy.

In building these machines, one has to supply electric power to each
accelerating section spread over extended distances. Several oscillating
power supplies, all synchronized to each other, are to be designed, built
and operated. The most ambitious project is to build a linear accelerator
of 500 GeV kinetic energy, dubbed the international linear collider.[17]

Oscillating field accelerators were almost simultaneously con-
ceived along with electrostatic accelerators. The principle here is that
when a particle of charge q goes through an accelerating field of magnitude V volts n times, it will gain a kinetic energy of nqV, an n-fold
gain compared to a single traversal. As an oscillating field changes

---

[15]See Section 8.12.

[16]The linear accelerator was invented in 1928 by Rolf Wideroe (1902–1996), a
Norwegian physicist working in Germany.

[17]International linear collider: A pair of linear accelerators producing electron and
positron beams, which will be arranged to undergo head-on collisions for particle
physics research purposes. If funds become available, this machine may be built for
use in the 2020s. For details, visit http://www.linearcollider.org/

sign from positive to negative and vice versa, the trick is to synchro-
nize particle passage in the fields such that particles see accelerating
fields while they are in the field space and ensure that they are in a
field-free region when the polarity of the field is of a decelerating sign.

We then require that the time it takes for a particle to travel between two accelerating gaps be equal to the time it takes for the field
to reverse its sign from positive to negative and vice versa. If f is the
frequency of the field, then the time t is equal to $1/2f$. In this time, a
particle with a speed of $v = \beta c$ travels a distance

$$d = \frac{\beta c}{2f} \tag{8.3}$$

As we see below, at each stage the speed of a particle increases
as its kinetic energy increases due to acceleration. Thus for the gap
between n and the $n + 1$th stage of acceleration,

$$d_n = \frac{\beta c}{2f} = \frac{\sqrt{\frac{2T}{mc^2}}}{2f} = \frac{\sqrt{\frac{2nqV}{mc^2}}}{2f} \tag{8.4}$$

In the above equation, for a given species of particles (i.e., m and q
fixed), with an accelerating potential V of frequency f, the value of
$n = 1, 2, 3, \ldots$ for the 1st, 2nd, 3rd, $\ldots$ stages of acceleration. Thus

$$d_1 = \frac{\sqrt{\frac{2qV}{mc^2}}}{2f} \tag{8.5}$$

$$
\begin{aligned}
d_2 &= \sqrt{2}d_1 = 1.414d_1 \\
d_3 &= \sqrt{3}d_1 = 1.732d_1 \\
&\text{and} \\
d_n &= \sqrt{n}d_1
\end{aligned}
$$

Thus, the accelerating gaps should be separated by drift spaces of
increasing lengths of 1, 1.414, 1.732, $\ldots$, $\sqrt{n}$ for the 1st, 2nd, 3rd,$\ldots$,
nth stage of acceleration, resulting in $T = nqV$ electron volts. This
assertion is valid for low energy heavy particles. For faster particles,
Einsteinian relativity is essential as the $\sqrt{n}$ approximation does not
hold.

Example 8.3

Wiederøe accelerated potassium (^{39}K) ions to 50 keV energy by employing a 25 kV field oscillating at a frequency of 1 MHz. The frequency is such that the field completes one full cycle in 1 μs. The particle sees accelerating potential for $1/2$ μs and we should let it drift in the second half of each microsecond.

From data tables, the mass of ^{39}K, $A = 38.964$ amu $= 36.29$ GeV/c^2.

Writing the mass in units of keV/c^2 and the kinetic energy in keV, we deduce the speed of these 25 keV ions as

$$\beta = \sqrt{\frac{2T}{mc^2}} = \sqrt{\frac{2 \times 25}{36.29 \times 10^6}} = 0.00117 \qquad (8.6)$$

So, the distance between the first and second gaps (d_1) for ^{39}K ions, subject to 1 MHz ($f = 10^6$ Hz), is

$$d_1 = \frac{\beta c}{2f} = \frac{0.00117 \cdot 3 \times 10^8}{2 \times 10^6} = 0.175 \text{ m} \qquad (8.7)$$

If Wiederøe were to try to accelerate electrons with the same field strength of opposite sign, the drift length would increase by the inverse square root of the ratio of masses:

$$
\begin{aligned}
\text{drift space for electron} \quad &= \quad \text{drift-space for } ^{39}\text{K ion} \\
&\quad \times \sqrt{\frac{\text{mass of } ^{39}\text{K ions}}{\text{mass of electron}}} \\
&= \quad \text{drift space for } ^{39}\text{K ion} \times \sqrt{\frac{36.29 \times 10^3}{0.511}} \\
&= \quad \text{drift space for } ^{39}\text{K ion} \times 136.18 \\
&= \quad 23.83 \text{ m}
\end{aligned}
$$

Clearly, it is quite a long drift space with Wiederøe's original setup.

This principle, without further development would have been of limited use. As we saw above, we need too large a drift space to accelerate particles. One notices that the drift space is inversely proportional to the frequency. So, higher frequencies will allow us to fit the machine in a smaller space. The World War II effort in radar communications led to the development of radiofrequency sources of much higher frequencies (GHz). Linear electron accelerators became possible only in the late 1940s. Currently, medical electron accelerators found in hospitals around the world are linear in structure.

8.7 Cyclotron

Cyclotrons are, perhaps, the most widely known particle accelerators. They have evolved since they were invented in the early 1930s.[18] The basic principle is again acceleration in an oscillating field. Instead of sending charged particles in a straight line and applying accelerating fields along their paths, we use the same accelerating field and send particles into it to increase kinetic energies each time the particles pass through the field region. The particles move in circular paths due to a constant magnetic field applied perpendicular to their velocity, a well known classical physics phenomenon. The accelerator design is based on three simple observations.

1. As we have seen before, a charged particle of momentum[19] p and electric charge q passing through a magnetic field B (in teslas) will be bent to follow a trajectory of radius R given as

$$p = 300 \times nBR \qquad (8.8)$$

In the above equation, p is expressed in units of MeV/c, and the

---

[18]Ernest Lawrence (1901–1958) was the inventor of the cyclotron. This accelerator was the predecessor of all modern circular machines. They have evolved from modest MeV energies to the large hadron collider of 7 TeV (7 million MeV) energy.

[19]Many cyclotrons operate at MeV energies. So, we resort to MeV and MeV/c units in this chapter. Thus $p = 300 \times qBR$ in MeV/c units, where B is in tesla, R is in meters and $q = ne$ is the charge of the ion.

charge of the particle is $q = ne$. Thus, $n = 1$ for protons, $n = 2$ for doubly charged ions (He^{++}, Li^{++}, etc.) or $n = 3$ for triply charged ions (Li^{+++}, C^{+++}, etc.) and R is in meters.

Also, for non-relativistic energies, the kinetic energy is

$$T = \frac{p^2}{2m} = \frac{300^2 (BR)^2}{2m} = \frac{9 \times 10^4 (BR)^2}{2m} \text{MeV} \qquad (8.9)$$

It is of interest to note that the product of the magnetic field B and the radius of the orbit determine the maximum attainable momentum and the kinetic energy of a particle. That is to say, a magnetic field $B = 1$ Telsa applied over 0.5 m will give the same momentum and kinetic energy as a $B = 0.5$ tesla field applied over a 1 m radius.

2. Another simple and elegant physics feature is that the angular velocity of a charged particle in a constant magnetic field is independent of its momentum for non-relativistic energies.

The frequency (f) of revolution of a particle mass m is given by [20]

$$f = \frac{p}{2\pi m R} = \frac{qB}{2\pi m} \qquad (8.10)$$

For non-relativistic particles ($v \ll c$), the mass of a particle is constant. Notice that the right hand side of the above equation does not depend on the momentum of the particle. Thus, the revolution frequency of a particle of known mass and charge in a fixed magnetic field is constant and it is known as the cyclotron frequency of the particle. As stated above, it does not depend on the speed of a particle. An increase in speed results in an increase in the radius of a particle trajectory and the non-relativistic particle spirals outwards while maintaining constant angular frequency.

In MeV/c units for momentum and electric charge as multiples of electron charges, we write

$$f = \frac{300 \times qBc}{2\pi m}$$

---

[20] A particle moving at an angular frequency f (number of revolutions) in a circular orbit of radius R traverses a distance of $2\pi R$ in one revolution. Thus the speed is $v = 2\pi R f$, momentum $p = mv = 2\pi R f m$ and kinetic energy $T = \frac{p^2}{2m} = 2\pi^2 R^2 f^2 2m$.

with mass expressed in energy units as a multiple of MeV/c^2. The frequency is inversely proportional to the particle's mass. Heavier particles have lower frequencies.

The table below lists the cyclotron frequency of some particles of interest.

Particle	Mass (MeV/c^2)	Charge # $n = q/e$	Cyclotron frequency for $B = 1$ Tesla	Period $T = 1/f$
Proton	938	1	15.27 MHz	65.5 nsec
Deuteron	1,875	1	7.64 MHz	131 nsec
Electron	0.511	−1	28 Ghz	36 psec

From the above table, we conclude that a proton, deuteron and electron complete one full revolution in 65.5 nsec, 131 nsec and 36 psec, respectively, for a 1 Tesla magnetic field. At non-relativistic energies, the frequency of oscillation increases with increasing magnetic field strength (B). At much higher speeds $(\gamma \gg 1)$, the mass also increases. The situation becomes more complicated.

As mentioned above, the product of $B \cdot R$ determines the maximum momentum and kinetic energy. However, cyclotron frequency is proportional to the magnetic field. A choice of a B field would also require that we optimize the frequency of the accelerating field for each species of particle.

3. The particle must enter the accelerating field region at times when the field is of the correct sign, i.e., a positive field for negatively charged particles and a negative field for positively charged particles. So we have to match the oscillating field frequency with the cyclotron frequency of the particle. We note the frequency is $f \propto \frac{B}{m}$. Thus, for fixed frequency, we vary the field strength in proportion to the particle's mass to achieve proper acceleration. The mass of the deuteron is $1,875/938 \sim 2$ amu, compared to protons of 1 amu mass.

A 2 Tesla magnetic field for deuterons will achieve the same frequency as that of protons with a 1 tesla field. Much lighter electrons are of revolution frequencies of 28 GHz in a 1 tesla magnetic field. Thus, electron accelerators work with radiofrequency fields of GHz frequencies with manageable magnetic fields, while protons and other ion machines work at a few MHz.

Example 8.4

A cyclotron was designed with varying B fields such that the minimum $B \cdot R = 0.5$ tesla·meter and the maximum $B \cdot R = 1$ tesla·meter.

We recognize that singly charged particles ($n = 1$) can be accelerated over the momentum range 150–300 MeV/c. This information is enough to calculate kinetic energies of particles. If the maximum orbit radius is 1 m, the B fields must vary between 0.5–1 tesla. From this information, we can calculate kinetic energies of particles and the frequency of accelerating fields.

In the simplest version, two D shaped electrodes with a gap between them are connected to an oscillating field. These electrodes are placed in a magnetic field of strength B. If we keep the frequency of the accelerating field constant, the radius of curvature r of a particle increases monotonously with increasing momentum and the particle spirals outward and can be extracted for various applications. In a fixed magnetic field strength B, extending over a radius R, the particle gains kinetic energy.

$$T = \frac{(2\pi f R)^2 m}{2} = \frac{(300 \times qBR)^2}{2m} = \frac{q^2}{m} \times \frac{(300 \times BR)^2}{2} \qquad (8.11)$$

It is important to note that the kinetic energy increases proportional to the square of magnetic field strength and the maximum radius of curvature (the size of the cyclotron). Thus we have to optimize the magnetic field strength, the maximum radius of curvature and the strength (voltage) of the accelerating electric field to achieve the desired kinetic energies.

Example 8.5

A radiofrequency field of 40 kV oscillating at 20 MHz is used to accelerate protons up to 10 MeV kinetic energies. As we saw above, the cyclotron frequency of a proton is 15.27 MHz for a 1 tesla magnetic field. Thus, for resonance conditions, one should apply

$$B = \frac{20}{15.27} = 1.31 \text{ tesla}$$

At a 40 kV accelerating field,

$$\# \text{ of turns } = \frac{\text{maximum kinetic energy}}{\text{gain of energy per turn}} = \frac{10,000}{40} = 250 \text{ turns}$$

At a 1.31 tesla field, the cyclotron frequency of a proton is $15.57 \times 1.31 = 20$ MHz. The period of revolution is $1/f = 50$ nsec.

The time required to accelerate protons to the maximum 10 MeV = time period of revolution × total number of revolutions = 50 nsec × 250 = 12.5 μs.

For economy of magnetic field reasons, one might opt to operate at a sub-harmonic such as a $1/2$ resonance condition of 0.65 tesla for a revolution frequency of 10 MHz. In this condition, the radiofrequency field undergoes two oscillations for each successive traversal of the particle in the gap. Again the particle makes 250 turns, but now it takes 25 μs to attain 10 MeV kinetic energy. We have been presenting a simple arrangement with a constant magnetic field satisfying the above equation to do the trick. This is a first approximation. It is only a first approximation since, in relativity, a particle's mass increases with increasing kinetic energy or momentum. So we really have to write the cyclotron frequency

$$f_c = \frac{0.3qBc}{2\pi m\gamma}$$

where γ is the Lorentz factor we discussed earlier. To remind ourselves, for low speeds, $\gamma \sim 1$. Early cyclotrons accelerating protons and other ions were able to produce particle beams of a few MeV energies with this simple machine.

8.7.1   K-Value of a Cyclotron

The cyclotron community and medical cyclotron companies specify a K-parameter to indicate the acceleration capabilities of a unit. Simply stated, the K-value of a cyclotron is the maximum kinetic energy a proton can attain in the unit. As we saw above, an ion of mass m_i and charge n electron units gains a kinetic energy

$$T_i = \frac{9 \times 10^4 \cdot n^2 \cdot (BR)^2}{2m_i} \text{ [MeV]} \tag{8.12}$$

Thus

$$K = \frac{9 \times 10^4 \cdot (BR)^2}{2m_p} \text{ [MeV]} \tag{8.13}$$

where m_p is the proton mass in MeV units.

Then the ratio of kinetic energy of an ion to that of a proton is

$$\frac{T_i}{K} = n^2 \frac{m_p}{m_i} \tag{8.14}$$

To a very good approximation, $T_i = K \times n^2/A$, where A is the mass number of the ion (4 for alpha, 12 for ^{12}C, 28 for ^{28}Si etc. ....)

Example 8.6

The Lawrence cyclotron accelerated protons to 1.2 MeV, with an accelerating field frequency of 3.5 MHz. The radius of the cyclotron was 11 inches in diameter (28 cm).

Thus K = 1.2 MeV for the Lawrence Cyclotron. We find that

$$BR = \sqrt{\frac{K \times 2 \times m_p}{9 \times 10^4}} = \sqrt{\frac{1.2 \times 2 \times 938.3}{9 \times 10^4}} = 0.16 \text{ T} \tag{8.15}$$

Or the magnetic field $B = 0.16 \text{ Tm}/0.14 \text{ m} = 1.14$ T = 11.4 kilogauss.

The cyclotron frequency of a proton is 15.27 MHz at a 1 T magnetic field. For the Lawrence cyclotron of $B = 1.14$ T, the proton's cyclotron frequency is 17.4 MHz. The cyclotron frequency of a proton is thus approximately 1.1 times the accelerating field frequency. Thus, though not perfect resonance, it was good enough to accelerate protons.

Don't forget that these are all non-relativistic cases. They are good for low energies and heavier ions. Be careful when you consider light ions or electrons.

As the particle energy increases, its characteristic frequency at a fixed B field decreases since γ increases. If we operate the system at a fixed B field for a fixed frequency oscillating field, say, at f, the particle will soon be out of phase with respect to the oscillating electric field, limiting the kinetic energy of a particle to small values compared to its mass.

Two solutions suggest themselves to circumvent this problem. Either increasing the magnetic fields for outer orbits or introducing a time dependent variation of the frequency of the electric field will achieve the goal. Most present-day cyclotrons use magnetic fields varying with changing orbit radius. The magnetic field profiles are organized in a pre-designed way such that particles move in curved paths to achieve better focusing of the beams. [21]

8.8 Microtron

Another scheme was presented[22] in which one can use one linear accelerator section several times. They were called microtrons. The modern version of this accelerator is a race-track microtron. We insert a linear accelerator section between two identical semi-circular magnets $M1$ and $M2$, each one of field strength B. Say a particle at the first acceleration has a momentum p_1 and that it follows a path of radius of curvature r_1 in magnet M_1. At the end of one semi-circular bending, it emerges along a line parallel to the linear accelerator with its direction of motion reversed. It then enters the magnet M_2, where it will be bent

---

[21]It is useful to recognize that a dipole magnet which bends the charged particle acts as a prism for visible light, bending and dispersing the light of different wavelengths in different directions. An electric quadrupole acts as an optical lens to focus or defocus the charged particles. The study and design of electric and magnetic fields is called "beam optics" to emphasize the similarity of the design of light transport systems in optical arrangements and charged particle transport.

[22]Vladimir Veksler (1907–1966), a Russian physicist, invented this machine in 1944. A brief review of his outstanding achievements can be found at http://cerncourier.com/cws/article/cern/30151/3

again with the same curvature r_1 to re-enter the accelerating section. If the field frequency is properly arranged, it will be accelerated again to emerge with increased momentum, say p_2. This particle will follow a path of radius r_2 and re-enter for acceleration to higher momentum p_3. This continues until a particle reaches a momentum p_n after n turns and then it is extracted from the accelerator for subsequent use.

Microtrons are mainly used to accelerate electrons.

Figure 8.4 is a schematic of a microtron. As shown, the accelerating field is well separated from the bending magnets. Inside the magnetic field regions, the particles follow semi-circular paths of increasing radii as energy increases. Outside the field region, they travel in straight lines.

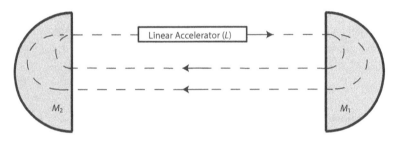

FIGURE 8.4: Schematic diagram of a microtron. Particles are accelerated in the linear accelerator L, operating at frequency ω. M_1 and M_2 are two semicircular magnets. The particles gain energy at each pass as they enter the accelerating region. They traverse semicircular paths of increasing radii as energy increases at each pass.

Example 8.7

The most modern microtron is at the Johannes Gutenberg University in Mainz, Germany. It is called the Mainzer Mikroton, abbreviated MaMi, and it accelerates electrons to 1.5 GeV energy. The MaMi facility is a complex of four accelerators (MaMi A1, MaMi A2, MaMi B and MaMi C).[23] This accelerator is one of the state-of-the-art machines of stable beams of electrons. It uses a DC gun of 100 keV energy to generate electrons. The electrons are accelerated to energies of

---

[23]For more information, visit http://wwwkph.kph.uni-mainz.de/B1//accelerator. php

3.97 MeV in a linear accelerator. The output electrons are injected into the microtron. In the first stage (A1), electrons are accelerated to 14.8 MeV in 18 recirculations. Then these electrons enter stage A2 where they are accelerated to 180 MeV after 51 recirculations, gaining 3.24 MeV per each turn. Electrons then enter the stage B, where they can be accelerated to 855 MeV after 90 recirculations. In the last stage C, they are recirculated 43 times (15 MeV gain per turn) to a final energy of up to 1.5 GeV. Each stage has its own linear accelerator to give the necessary boost in the energy of the particles. This accelerator can deliver high intensity beams of electrons of up to 100 μA (6×10^{14} electrons/second).

8.9 Synchrotron

In a cyclotron or a microtron, particles travel in orbits of increasing radii with increasing momenta. In these machines, we keep the magnetic field B constant, resulting in an increase of radius R for increased momenta. A synchrotron is an arrangement in which the radius R is kept constant and the magnetic field B increases in sync with the increase in particle energy. These days, high energy accelerators are built on this principle.

At first look, synchronization of particle energies and magnetic fields may be a very difficult task. However, physics is very forgiving. McMillan[24] proposed an ingenious phase stability principle which is exploited heavily in modern accelerators.

As described before, it is the relative phase of the accelerating field and arrival of the particle which needs to be synchronized. Assume we work with a fixed accelerating field frequency. Adjust the B field such that the particles of equilibrium energy with Lorentz factor $\gamma = \gamma_e$ arrive at the accelerating gap when the field amplitude is zero so that they gain no kinetic energy. If the particle has higher than the equilibrium energy ($\gamma > \gamma_e$), the angular frequency decreases (f is proportional to

---

[24]Edwin M. McMillan, *Physical Review*, vol. 68, 1945, page 143.

$^1/_\gamma$) and it comes later than those particles of equilibrium energy. On the other hand, particles with less than equilibrium energy have greater angular frequency and they come earlier than those with equilibrium energy. So arrange the equilibrium particle to arrive at a phase angle greater than 90 degrees, on the decreasing side of the accelerating field amplitude. Then the particles of higher than equilibrium energy gain less energy and get closer to equilibrium energy. The particles with lower than equilibrium energy will gain energy to be closer to equilibrium energy. Simply stated, the accelerating phase is such that a particle which arrives early in one turn (higher energy) will arrive later in the subsequent turn. The ones which arrive late in one turn come early in the following turn. In a few turns all the particles are grouped together at the equilibrium energy. We then increase the magnetic field to a higher value, thus increasing the cyclotron frequency of particles. Now the particles will arrive earlier, experience an accelerating field and their energy increases to a new equilibrium value. As the energy increases, the Lorentz factor γ of the particles increases, resulting in a decrease of cyclotron frequency, gradually settling to the equilibrium value again. We repeat this process of increasing the magnetic fields slowly to the optimum value, thus increasing particle energies at each step. All modern synchrotrons make use of this principle to accelerate particles to higher and higher energies.

Synchrotron light sources are based on electron acceleration to high energies and let the electrons circulate in a storage ring kept under high vacuum. As electrons are bent from a straight path, they emit synchrotron radiation. The energy spectrum of synchrotron radiation depends on the bending path of electrons. Present-day synchrotrons offer several exit ports for synchrotron radiation, each designed to optimize the energy spectrum of emitted photons to suit specific applications.[25]

---

[25]A list of worldwide light sources can be found at http://www. lightsources.org/cms/?pid=1000098 with links to the websites of those facilities. Similar to research nuclear reactors, the light source facilities accommodate several users, each with unique experimental programs and diverse needs for photon beam properties. Thus they are workhorses for material science, engineering and several biological, chemical and other research areas.

8.10 Collider Machines

As we have seen in the chapter on energetics, when a beam interacts with a stationary target nucleus, the energy required to cause a reaction with a negative Q-value is

$$E_{\text{threshold}} = \frac{m_a + m_b}{m_b} |Q| \tag{8.16}$$

Here $E_{\text{threshold}}$ is the minimum kinetic energy of the beam particles; m_a and m_b are the masses of the stationary target particle and beam particle, respectively. Clearly, $E_{\text{threshold}}$ is higher than the magnitude of the Q-value. As we saw earlier, this excess energy is required to satisfy the energy and momentum conservation laws.

In a stationary target arrangement, the target is at rest and is of zero momentum. The beam particles have a finite momentum. Thus it is essential that the final state products must also carry the same momentum to satisfy momentum conservation. The energy available for physics processes is the center of mass energy.

The center of mass energy can be written as

$$E_{cm} = \left[m_a^2 + m_b^2 + 2 \times m_a (m_b + T) \right]^{1/2} \tag{8.17}$$

Here m_a and m_b are the masses of the target and beam particles in energy units and T is the kinetic energy of the beam particle. For identical particles of beam and target, we may write $m_a = m_b = m$. We then have

$$E_{cm} = \sqrt{2m} \left[2m + T \right]^{1/2} \tag{8.18}$$

For very high kinetic energies, $T \gg m$, we approximate as

$$E_{cm} \approx \sqrt{2mT} \tag{8.19}$$

That is, the available center of mass energy increases as the square root of the kinetic energy. For example, a fourfold increase in kinetic energy increases the center of mass energy by a factor of two.

What if we arrange the experiment such that the particles before they interact have zero total momentum, i.e., if two particles undergo head-on collisions with equal and opposite momenta? Then the final state particles will also be of zero momentum. They are distributed in the laboratory such that the final vector sum of all momenta is zero before and after the collisions. In this situation, the kinetic energy required is simply the Q-value for the process to occur. For the above example, if two identical particles of the same mass m and the same kinetic energy T collide head-on, then the center of mass energy is

$$E_{cm} = 2(m + T) \tag{8.20}$$

The center of mass energy increases linearly with the kinetic energy of the particles.

Several collider facilities, mostly for particle physics research, were built in the world. In the USA, Brookhaven National Laboratories in Long Island, New York; Fermi Lab near Chicago; Stanford Laboratory are examples. There are laboratories in Europe, Japan and China. We mentioned the linear collider facility in the planning stages.

Before we close this section, it may be of interest to introduce the concept of the luminosity of a collider beam facility.

In particle accelerators, we specify the beam current as a measure of intensities. For example, for singly charged particle beams such as protons or electrons, we specify the subunits of amperes as intensity measures. An electron or proton or any singly charged particle carries a charge of $q = e = 1.6 \times 10^{-19}$ C. Electric current is said to be 1 A if 1 C of charge flows per second. Thus 1 A current of electron, proton, etc., beams corresponds to the number of particles (n):

$$n \times e = 1\frac{C}{s}$$

or 1 A:

$$n = \frac{1}{1.6 \times 10^{-19}} \frac{C/s}{C} = 6.25 \times 10^{18} \text{particles/s}$$

Clearly, doubly charged particles, such as alphas, C^{++}, etc., will be 1 A = 3.125×10^{18} particles/s.

Medical accelerators produce, say, proton beams of about 1 mA, which corresponds to 6.25×10^{15} protons/s. Usually, nuclear or particle physics experiments are done with beams of a few microamperes or even nanoamperes. In particle colliders, the fluxes and overlap areas of the colliding beams are measures of interactions. This is expressed by luminosity. See the appendix for details.

8.11 Questions

1. An ion beam facility for material science comprises a tandem van de Graaff accelerator of terminal voltage 1.7 million volts. It consists of a source of singly charged negative ions, which are directed to the terminal. At the terminal, the ions are stripped of a few electrons to produce positive ions of 1, 2, 3 ... charges. They are then further accelerated to deliver beams for the experiments. We accelerate singly charged hydrogen, doubly charged helium (alpha), triply charged carbon ions and silicon ions of charge 4. What are the kinetic energies of these ions?

   All these ions must be bent to be of the same radius of curvature to arrive at the experimental target.

   Say we need to employ a 0.5 T magnetic field to bend the protons (hydrogen ions) to meet this need. What are the corresponding magnetic fields for the alphas, carbon and silicon ions mentioned above?

2. In the discussion of cyclotrons, we mainly focuesd on ions (protons, alphas, etc.) but ignored electron acceleration. In fact, cyclotrons are not used to produce MeV electrons. Can you guess why not?

   To convince yourself about the complexities, consider accelerating electrons to a kinetic energy of, say, 50 MeV in a magnetic field to be applied over a radius of 50 cm ($R = 0.5$ m).

   Calculate the radii of orbits and cyclotron frequencies of electrons of 1, 5, 10, 20 and 50 MeV kinetic energies, taking into account the change in electron mass with the kinetic energy. Hence, deduce the magnetic field strength versus distance from the middle (source point) of the cyclotron. Can you see the difficulties? What are they?

3. In the Large Hadron Collider (LHC), two proton beams, each of 7

TeV energy, undergo head-on collisions. What is the center of mass energy of these collisions?

If we were to build an accelerator with high energy protons colliding on a hydrogen gas target (protons at rest in the laboratory), what should be energy of accelerated protons to achieve the same center of mass energy as at the LHC?

8.12 Endnotes

Footnote 6

The Lorentz force with only an electrostatic field is

$$\vec{F} = e\vec{E} = e\frac{V}{d}\hat{z} = m\vec{a} = ma\hat{z}$$

where e and m are the charge and mass of the electron d is the distance between the cathode and the anode. Then the electric field, a vector quantity, is given by

$$\vec{E} = \frac{V}{d}\hat{z}$$

$\hat{z}$ is the field along a straight line connecting the centers of the cathode and the anode, the direction along which the electron will be accelerated.

We may then write the magnitude of acceleration a as

$$a = \frac{eV}{md}$$

It is interesting to note that this understanding is enough to determine the motion of the electron completely, its velocity, its time of transit and acceleration, etc.

Within Newtonian mechanics, we have equations connecting the velocities v, acceleration a, distance traveled d and the time of transit in the accelerating field t_d as

$$
\begin{aligned}
v^2 &= u^2 + 2ad \\
v &= u + at_d \\
d &= ut_d + \frac{1}{2}at_d^2
\end{aligned}
$$

where u and v are the initial and final velocities of the particle as it enters (u) and leaves (v) the field region.

We can calculate the time t_d it takes for an electron to travel from the cathode to the anode to be

$$t_d = \sqrt{\frac{2d}{a}} = \sqrt{\frac{2md^2}{eV}}$$

While in transit, the speed of the electron v_s at any point s it reaches in time t_s is given from the relations

$$v_s = a \times t_s \text{ and } s = \frac{1}{2}a \times t_s^2$$

as $v_s = \sqrt{2as}$.

The increase in the kinetic energy dT of a particle as its speed increases from v to $v + dv$ is given by

$$dT = m \cdot v \cdot dv = m \cdot at \cdot d(at) = ma^2 t dt$$

for constant acceleration. The kinetic energy after traveling a distance d in time t_d is

$$T = \int_0^{t_d} ma^2 t dt = \frac{ma^2 t_d^2}{2} = \frac{m}{2} \times \left[\frac{2md^2}{eV} \right] \times \left[\frac{eV}{md} \right]^2 = eV$$

Example 8.8

A commercial X-ray tube is of the following design parameters:

Cathode-anode potential difference = 100 kV; cathode-anode distance = 15 cm. The electric field is

$$\frac{100 \text{ kV}}{0.15 \text{ m}} = 6.7 \times 10^5 \text{ V} \cdot (\text{m}^{-1})$$

The acceleration is

$$\frac{eV}{md} = \frac{100 \text{ kV}}{mc^2} \times \frac{c^2}{0.15 \text{ m}} = \frac{10^5}{0.511 \times 10^6} \times \frac{(3 \times 10^8)^2}{0.15} \text{ms}^{-2}$$
$$= 1.17 \times 10^{17} \text{ ms}^{-2}$$

This number might look too big to be physically realistic, but it acts on a very light particle for a very short period of time.

$$t_d = \sqrt{\frac{2md^2}{eV}} = \sqrt{\frac{2 \times mc^2 d^2}{10^5 c^2}} = \sqrt{\frac{2 \times 0.511 \times (0.15)^2}{10^5 \times 9 \times 10^{16}}} \, s = 1.6 \times 10^{-9} \, s$$

This is just less than a 1 nanosecond time interval.

Note that we employed non-relativistic (Newtonian) kinematic relations for this problem. As kinetic energy of 100 keV is much less than the rest energy of an electron (511 keV), the numbers are fairly accurate for the speed and time estimates.

Footnote 11

This principle was discovered by Heinrich Greinacher (1880–1974). It is based on the principle of the transformer, which either steps up (increases) or steps down (decreases) oscillating voltages. This feature, combined with the characteristic of electric diodes that they allow currents to pass only in one direction, requiring that the anode be positive with respect to the cathode and that the capacitors are charge storage devices, was ingeniously used by Greinacher to build voltage doubler circuits. This principle was soon extended to many-fold multiplication of the voltage.

Footnote 14

An accelerating potential V acts on an electric charge q, say, n times; the kinetic energy of the particle is $T = nqV = (\gamma - 1)mc^2$. Or

$$\beta = \frac{\sqrt{T^2 + 2mc^2 T}}{T + mc^2} = \frac{\sqrt{1 + \frac{2mc^2}{T}}}{1 + \frac{mc^2}{T}}$$

For low velocities, $T \ll mc^2$, we can neglect 1 in both numerator and denominator to get

$$\beta = \sqrt{\frac{2T}{mc^2}} = \sqrt{\frac{2nqV}{mc^2}}$$

This equation is the Newtonian limit for low energies. The speed of a particle of mass m increases as $\sqrt{1}, \sqrt{2}, \sqrt{3}, \ldots, \sqrt{n} = 1, 1.414, 1.732, \ldots, \sqrt{n}$ for the 1st, 2nd, 3rd, $\ldots$, n stages of acceleration.

Footnote 15

A general analogy is with ocean surfing. Ocean surfers are instructed to watch the waves carefully and catch the waves when a wave is rising and not when it is breaking. They should catch a wave such that as they start to stand the wave should be breaking under them. If the wave is breaking before they stand, they are a bit too late to catch the wave. If the wave is moving and breaking later, they are a bit too soon. Thus the surfers are synchronizing their actions with the phase of ocean waves. Most of the time, surfers cannot accurately predict when a wave will break, but electromagnetic field waves are quite predictable and thus the functioning of a well built accelerator is not erratic.

9

Nuclear Radiation Detectors

9.1 Introduction

The previous chapter was concerned with techniques of producing particle beams of various species. The beams are then put to use for scientific, medical or industrial uses. Understandably, interactions of radiations with material media as described in the first sections influence the outcomes of applications. Not surprisingly, the same physics of interaction of radiations and material properties which governs the radiation phenomena is applicable for the development of radiation detectors. The detectors are transducers, converting ionizations or energy deposits of radiations to electrical or optical signals. The signals are processed to obtain the information we seek, which may be

(i) radiation levels in a nuclear facility such as a particle accelerator or a nuclear reactor or for environmental monitoring

(ii) energies and intensities of various species of radiations for measurements in research, teaching, medical or other applications

(iii) identification and quantification of various species of radiations in a complex radiation environment for nuclear or particle science purposes

(iv) energy and time correlations of radiation species for research and applications

The detector systems evolved from visual systems (counting fluorescence signals with the naked eye) and slow photographic systems to high speed electronic arrangements with associated digitization and graphic representations. As technologies change rapidly, detector instrumentation is also advancing in leaps and bounds. Thus the discussion below is not exhaustive but it is intended to provide a flavor of various techniques and the physics behind them. A good resource to keep up to date with on-going developments of particle detectors is the review of particle physics[1] published biennially.

The aim is to generate electrical signals to derive necessary information about radiation. An interaction of radiation with matter may release either a single photon or an electron, which may not be directly observed. The task is to amplify it to measurable levels. One has to design these systems by taking advantage of the electrical and optical properties of materials and new electronic and digital technologies.

The arrangement can be a simple detector such as a Geiger–Müller counter measuring radiation levels or a multi-detector assembly for medical imaging by positron emission tomography. It may be a highly sophisticated ATLAS detector[2] at the CERN in Switzerland which analyzes complex data of millions of particles per second or anything in between. The underlying principles are based on the physics of interactions of radiation with matter, techniques to derive useful electrical signals, further electronic processing and computer systems for acquiring and analyzing data. The complexity may vary widely, but the physics principles remain the same. In this chapter, we focus on various transducers useful for each type of radiation and their principles.

One of the earliest radiation detectors was a zinc sulfide (ZnS) phosphor, used by Rutherford and his collaborators to study the properties of alpha particles and the discovery of the atomic nucleus. ZnS phosphor emits light when electrons or alpha particles are incident on

---

[1]Visit http://pdg.lbl.gov/2012/reviews/contents_sports.html and click on the link Experimental Methods and Colliders to download PDF files.

[2]Please visit http://www.atlas.ch/detector.html. Briefly, it consists of four major components, each composed of several hundreds or thousands of carefully designed, built and operated detector elements. Simply stated, this detector is equivalent to a composite system of several million detectors working in concert to cope with data of about 1 billion events per second. The purpose is to measure physical parameters of energies, momenta, velocities and types of radiations in each recorded event.

it.[3] It was a predecessor to modern scintillation counters which employ photomultipliers or photo avalanche diodes to generate fast electrical signals with time resolutions of nanoseconds or less and count rates of thousands of events per second. Advances in material sciences, electronics and computer systems, all of which have undergone revolutionary changes, are contributing to these developments.

Generally speaking there are two types of detectors:

(i) ionization detectors which create charge carriers and

(ii) optical sensors which create photons as the result of nuclear radiations passing through them.

Gas detectors and solid state detectors generate electrical signals which can be processed by electronic circuitry. Scintillation counters and Cherenkov detectors generate optical signals, which must be converted to electric pulses by photon detectors before they can be input for electronic manipulation.

Based on the application, we may classify the detectors as calorimetric systems, tracking detectors or hit detectors. Calorimetric systems are designed such that they absorb the energy of radiation for energy spectroscopy purposes. Hit detectors register the passage of radiation through them for timing purposes or generate a logic signal to validate the data.

In general, a tracking detector is a series of hit detectors registering the passage of a single particle. Tracking detectors operate on a simple principle. A charged particle ionizes the medium along its path. We can reconstruct the trajectory of a particle by sampling ionization along its path. The ionization creates both negatively charged electrons and pos-

---

[3]The most noteworthy early experiment was that of Rutherford and his colleagues for the discovery of the atomic nucleus. They used a collimated alpha particle source, sending a stream of alphas onto a gold foil. A thin ZnS film mounted on a microscope could be swivelled around an axis with a gold foil at the center. With this arrangement, Rutherford's colleagues could measure as many as 100 alpha particles per minute, limited by the speed of response and the sensitivity of the human eye. They measured intensities of alpha particles (number alphas detected, say, per minute), positioning the microscopic axis at several fixed angles with respect to the collimator.

itively charged ions. The earliest track detectors were cloud chambers[4] and bubble chambers.[5]

9.2   Cloud Chambers and Bubble Chambers

They make use of bubble formation in supersaturated and superheated vapors to track a charged particle. As particles pass through vapors in such chambers, condensation occurs along their path. Photographic techniques were used to record the information. Cloud chambers were extensively used in the study of cosmic rays until about the 1940s. Notable experiments include the discovery of the positron[6] (anti-electron) and the muon.[7]

A cloud chamber relies on the principle that a supersaturated liquid vapor is in a metastable condition. A charged particle passing through it causes ionization, resulting in condensation along its path. A cloud chamber can be easily made for demonstration purposes. A good website for the cloud chamber demonstration is Jefferson Laboratory's science education.[8]

A bubble chamber is maintained in a superheated state and the pressure is slowly reduced just before particles enter it. By reducing pressure, we render the liquid unstable and more prone to boiling. Bubble chambers rely on the physics principle that the boiling point of a liquid decreases at lower pressures. A charged particle ionizes the liquid medium, resulting in an increase of temperature and local boiling along its path. Pictures of the bubbles will reveal the particle trajectory. This is analogous to the way we see the path of a jet plane in the sky from the trail it leaves behind. We can also apply a known magnetic field across the chamber to make charged particles follow curved paths. From the sign of the curvature (concave or convex) and the radius of curvature

---

[4]C. T. R. Wilson (1869–1959) invented the cloud chamber in 1911.

[5]Donald Glaser (1926–) invented the bubble chamber in 1952.

[6]Carl D. Anderson in Physical Review, Vol. 43, pages 491–494, 1933.

[7]J.C. Street and E.C. Stevenson in Physical Review, Vol. 52, pages 1003–04, 1937.

[8]Visit http://education.jlab.org/frost/cloud_chamber.html

of the bent path, we can determine if the particle is negatively or positively charged and its momentum.

Liquid bubble chambers were mainly used for particle physics in the need to study the interaction of protons with some beam particles. Thus they were employing liquid hydrogen. Liquid hydrogen boils at 20 K ($-253°$C) at normal pressure. The idea is to prepare a bubble-free liquid hydrogen under high pressure at high temperature and have it be sensitive enough to create bubbles when charged particles move in it. To this end, a liquid hydrogen chamber at 30 K and at a pressure of 5 bars is prepared. Then the pressure is decreased to 2 bars just before a particle beam from the accelerator enters the chamber. This reduction in pressure increases the sensitivity for bubble formation as particles pass through the chamber. Several secondary particles may be produced as a result of interactions of beam particles with protons of hydrogen molecules. The tracks of primary beam particles and secondary particles are photographed for later analyses. The physics information extracted from the track analysis for their curvatures and signs were heavily exploited by the particle physics community for making important discoveries. A notable example is the discovery an exotic particle known as the Ω-baryon at the Brookhaven Laboratory, Long Island, NY.[9]

9.3 Gas Detectors

Another development was gas chambers in which highly energetic particles leave trails of ionization. Gas detectors are used for many applications. They are based on the principle that a charged particle, moving in a gas medium, releases electrons as it undergoes collisions with molecules. In the absence of external fields, electrons are again attracted to positive ions; they recombine and we see no net effect.

However, if we subject the gas medium to electric fields, two things happen.

1. Electrons drift toward the anode (positive electrode). Along the way, they gain energy and they can liberate more electrons due to collisions with gas molecules.

---

[9]Visit http://www.bnl.gov/bnlweb/history/Omega-minus.asp

2. The secondary electrons, in turn, gain energy and they cause further ionization and so forth.

The above two phenomena depend very much on the gas pressure and electric field strengths. The task is to arrange electric field profiles such that charges arrive at an electrode, signaling the passage of a particle. As electrons are much lighter than ions (the mass of an electron is less than 1/1836 times that of a proton, the lightest ion), they move swiftly and are captured faster at the electrode. They are thus better time markers than positive ions.

This type of detector is a simple arrangement of two electrodes in a gas chamber. A small sealed tube contains an anode at a positive voltage to which electrons are drawn. The chamber wall at ground potential (zero voltage) serves as the cathode. They come in various sizes and geometrical shapes.

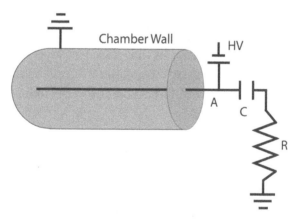

FIGURE 9.1: Sketch of a gas counter. The chamber wall is at ground potential. A is the anode wire held at high voltage (HV), C is the blocking capacitor and R is the load resistance.

Figure 9.1 is a sketch of a simple gas chamber, consisting of an anode wire and ground as a cathode. Electrons caused by ionization drift to the anode wire and they are collected at the anode, rendering the anode slightly less positive. As electrons are being collected, there is a change in electric current through the anode. The capacitor C lets this fluctuation pass across the resistor R, which appears as a voltage pulse. Thus the appearance of a voltage pulse across the resistor signals that a particle has been successfully detected by the gas detector.

In a gas chamber at fixed pressure, gas ionization exhibits distinct characteristics as we vary the voltage or field strength.[10] For radiation detection, three distinct modes of operation are possible and they are described below.

9.3.1 Ionization Chambers

At very low voltages, electrons ionized by radiation do not have enough energy and they may recombine and be lost. As we increase voltage, some energetic electrons reach the anode and a signal is generated. With further increases of voltage, more and more primary electrons (electrons produced by the passing radiation) will be collected at the anode, resulting in bigger signals. For a range of voltages, known as the ionization region, all primary electrons are collected. The collected charge is still too small to count individual particles of incident radiation. A common application is to measure the electric current to determine the average fluxes of incident radiation. With proper calibration, one can then measure radiation levels in environmental or radiation facilities such as particle accelerators or nuclear reactors. They serve as diagnostic tools to monitor the performance of these machines.

A common household use of ionization chambers is for smoke detectors in buildings. They are free air ionization chambers. In this chamber, a source emitting alpha radiation is used to create a constant ionization in the chamber to be collected at the anode. In ambient air circulation conditions, a constant current flows through the chamber. When smoke enters the detector, the ionization charges recombine with the smoke particles, resulting in less ionization current. This decrease in ionization current is made to trigger an alarm sound. The ^{241}Am isotope (half-life of 432 years), emitting alpha particles of 5.48 MeV energy, is most commonly used for this purpose.

The mean ionization potential[11] of air is 34 V. Thus, a 5.48 MeV

---

[10]Remember that electric field E = voltage/distance between electrodes.

[11]Ionization energy is the energy (expressed in eV units) required to free an electron from its atoms or molecules. Alternately, it is the change in the potential energy of an electron for it to be free from the bound state. Ionization potential is numerically equal to the ionization energy, expressed in volts. In the above example, an electron in an air molecule can be freed if it receives an energy of 34 eV. Thus the ionization energy is 34 eV and the ionization potential is 34 V.

alpha particle emitted by an ^{241}Am source can create $5.48 \times 10^6/34 \sim$ 160,000 ion pairs. While this might look like a big number, the current flow is only about 2.56×10^{-14} A[12] for 1 Bq of ^{241}Am. Domestic smoke detectors may be fitted with a source of 1 μC_i (37,000 Bq), inducing a current of 9.47×10^{-10} A or just about 1 nA.

Another common use is to monitor nuclear power or research reactors. As we know, nuclear reactors operate on nuclear fission and they are a source of high neutron fluxes. There are different stages of monitoring a reactor. As we start up a reactor, neutron flux increases, stays nearly constant during steady operation and decreases as we shut down the reactor. The task is to monitor each stage to ensure that the reactor is not behaving erratically. Besides monitoring temperatures, pressures etc., a direct way is to measure the neutron flux changes. Neutrons do not cause direct ionization. However, they cause fission (basic physics principle of a nuclear reactor), releasing heavy charged ions. They are also very easily absorbed by a ^{10}B nucleus, resulting in the emission of alpha particles via the process

$$^{10}B + n \rightarrow ^7Li + \alpha$$

In either case, a nuclear process results in highly ionizing charged particles, which can be easily detected by the ionization chamber. To this end, the walls are coated with fissionable material (^{235}U, ^{238}U or ^{233}Th) for fission chambers and with ^{10}B for boron ionization chambers.

9.3.2   Proportional Counter

If we increase the voltage to values beyond the ionization region, electrons of the primary ionization gain enough energy to cause further ionization, liberating more electrons. Understandably, the highest energies are in the proximity of the anode just before the electrons are collected. With increasing voltages, the secondary ionization increases and thus the electric current. Proportional counters operate in the region where each primary electron creates at most one secondary electron. The voltage pulse thus generated is proportional to the amount of primary ionization and so these are called proportional counters. They

---

[12]This is because one electron carries an electric charge of 1.6×10^{-19} C.

can be used to distinguish highly ionizing alpha particles from betas and photons.

9.3.3 Geiger–Müller Counter

At voltages higher than the proportional region, there is a range for which the gas chamber is nearly fully ionized. This is where a charged particle produces detectable ionization, independent of its energy. In this region, the detectors are sensitive to gamma rays also. Electrons liberated by the photoelectric effect or Compton scattering can cause enough ionization to be registered by the detector. It thus proves to be a very efficient, inexpensive detector to measure radiation levels. No wonder it is in use after more than a century after its invention. It is not recommended to increase the voltage beyond the Geiger region. Such an action will lead to the breakdown of the gas and irreparable damage to the chamber.

9.4 Multi-Wire Chambers

A single wire gas detector, as above, provides limited information of radiation passing through it in a given time interval. The spatial resolution is limited to a few millimeters, the size of the detector. The time resolutions are about a few hundred nanoseconds or longer. When we need to collect radiations passing through large areas or several particles flying in the laboratory, we need to stack up a lot of these detectors and build electronics around them. In the last few decades, great progress has been made in the development of multi-wire chambers[13] to improve efficiency, enhancing spatial and temporal resolutions and also multi-particle detection capability. A multi-wire chamber is simply several anode wires strung inside one large chamber, thus minimizing dead space and reducing hardware.

9.4.1 Multi-Wire Proportional Chambers (MWPC)

This is a simple construction of a few equally spaced anode wires in a chamber with two chamber walls serving as cathodes. Usually, a sin-

---

[13]Groups at CERN, under the leadership of Georges Charpak (1924–2010), were the main driving forces in the development of these detectors.

gle power supply can drive all the anodes connected in parallel. Each anode wire is connected to electronics just as in the case of single wire chambers. The outputs are fed into counters which will register the number of events recorded by each wire. The spacing between two neighboring anodes can be as small as a couple of millimeters, thus affording spatial resolution of about a millimeter.

We guide several charged particles through a magnetic field (B tesla) such that each particle of momentum (p GeV/c) follows a trajectory of radius of curvature R meters. As we have seen before, the equation

$$p = 0.3 \times n \times B \times R \tag{9.1}$$

connects these parameters. Here, n is the charge state of the ion, i.e., $n = 1, 2$, for singly and doubly charged ions. If we now place a set of wire chambers (WC) at the entrance and exit of the magnetic field, each anode in the chamber corresponds to a unique R and the relation for each arrangement can be determined by calculations and confirmed by calibrations. Thus, in an experiment, we can determine the momentum spectra of emissions by simply registering the number of counts recorded at each anode wire. While this arrangement can distinguish positive charges from negative ones, it cannot discern different particle species of the same momentum. For example, protons, deuterons, tritions or singly charged ^{3}He and ^{4}He ions, all of the same momentum, will induce signals on the same anode and they are not distinguished by wire chamber information only. We need additional data which is sensitive to the speeds of particles to identify particle species.

9.4.2   Multi-Wire Drift Chambers (MWDC)

This is the next development. In this arrangement, electric field profiles in the chamber are optimized by introducing field shaping electrodes between anodes. It allows for measurements of times of arrival of secondary ionization as electrons drift to an anode. One can then collect signals from a few neighboring anodes and relative times of arrivals of signals and then reconstruct the distances of origin of ionization (path of primary radiation) in the chamber with respect to anodes carrying signals. With fast electronics and software analysis, one can achieve position resolutions of about 100 μm for anode spacings of 5–10 mm.

In comparison to MWPC, better than a factor of 10 improvement is achieved with fewer anodes and associated electronics. As secondary ionization is proportional to energy deposits of primary radiation, arrangements to collect the charges arriving at cathodes also have been implemented.

It is of interest to note that in general operating conditions with many commonly used gases, the drift speed of electrons (v_d) is $v_d = 1$ mm per 20 ns. Thus from simple geometry of the spacing between the anode layer and the cathode plane, we can estimate the detector deadtime.[14] For example, a 5 mm spacing causes a deadtime of about 100 ns.

More recent developments based on the same principle of gas ionization are micro strip gas chambers, gas electron multiplier (GEM) and micro-mesh gaseous structure (micromegas) detectors. They avoid the use of wires for anodes. The first among them was the micro strip gas chamber.[15] It consists of a resistive plate on to which tiny metal strips are engraved. Alternate strips serve as anodes and cathodes, limiting the avalanches to narrow spaces. This idea was further developed in GEM and micromegas detectors to withstand very high radiation environments. They achieve high spatial resolutions to locate the particle paths to a few microns precision and they withstand high count rates.

The important thing to recognize is that in all these detectors, the basic physics principle is the same. A primary particle causes ionization in which negative lighter electrons move fast and are collected at the anode. By collecting the ionization, we can determine that a particle has passed through the detector. We may also localize its path better by registering time information. Then we are concerned about how to make a sturdy detector which is easy to operate with the least amount of electronics, able to withstand high levels of radiation and has a long life. But the physics of detection is unchanged.

These tracking detectors, while they are effective for particle tracking and thus momentum determination by magnetic fields, are not always the best or the simplest solution for a task at hand.

---

[14]The deadtime of a detector is the minimum time interval between the arrival times of two particles, below which the detector sees them as one single event.

[15]For a good review, see Fabio Sauli and Archana Sharma in Annual Review of Nuclear and Particle Science, vol. 49, pages 341–388, 1999.

For example, if we have a few emissions of low energies of alphas, betas or gamma rays, a compact detector with a few electronics can serve the needs. Scintillation and solid state detectors have been in use for these purposes. Below, we discuss the physics of these transducers.

9.5   Semiconductor Solid State Detectors

Semiconductors have been in use as radiation detectors for about half a century, soon after their invention. Semiconductor detectors are based on ionization and reverse bias properties. In comparison to gas chambers, the advantages of semiconductors are that they are

(i) compact in size

(ii) operate at low voltages

(iii) highly sensitive

(iv) time response faster than gas ionization detectors

(v) energy loss information useful for spectroscopy, etc., is available

Almost all semiconductor solid state detectors are based on the reverse biased diode principle. The diodes are basic elements of many electronic and computer systems. A diode is said to be forward biased when the anode is positive with respect to the cathode and it readily conducts electricity in this mode. When the anode is negative with respect to the cathode, i.e., reverse biased, it does not normally conduct electricity. However, energy loss by nuclear radiation can liberate the charge carriers to result in electric current. The number of charge carriers is proportional to the energy loss of the particle and this lends to energy spectroscopy.

Silicon and germanium crystals are the most popular semiconductor detector materials. More recently, other combinations such as cadmium-tellurium (CdTe) have been used. When it comes to gamma ray energy spectroscopy of several tens of keV to a few MeV, we look for a compact system in which photons will likely deposit all their energy. This is done by the photoelectric effect or pair production as the physical mechanisms. Both these processes are dominant in high Z media. That is the reason germanium (Z = 32), cadmium (Z = 48) and tellurium (Z = 52) with good physical properties are commonly employed.

It is not that solid state detectors render gas chambers obsolete, old technologies. Gas detectors are still better and the only viable choice for certain applications. For example, they are quite economical for large area detectors, while still achieving good position resolution of particle tracks. Gas chambers can be operated at normal STP conditions. Also, we can draw signals from gas chambers without the particles losing too much energy in generating signals. For some applications with relativistic particles in a magnetic fields, gas chambers and a thin tracking detector strategically spaced along the tracks of particles are desired. In general, some combinations of gas and solid strip detectors are usually employed.

9.6 Scintillation Counters

The gas chambers and solid state detectors described above make use of the ionization properties of radiation. Scintillation counters rely on light emitting properties of phosphors. This is a two-step process. First, we need a material which generates scintillations as radiation deposits energy in it. Second, we need a photon sensor which converts the scintillations into an electrical signal.

Scintillation materials are found in gas, liquid and solid states. A lot of research is still being done in search of high efficiency, robust, less expensive scintillating materials and configurations.

When passing through a fluorescence or phosphorescence[16] medium, charged particles cause the medium to emit scintillations, the intensity being proportional to the energy deposits by the particles. Above, we noted that ZnS is a phosphor detector with scintillations

---

[16]In reality, the difference between a fluorescent and phosphorescent medium is simply the difference in time scales involved between the passage of primary radiation and scintillation emissions. Florescence is when the time delay is negligibly small and phosphorescence has finite delays. Phosphorescence is commonly used for glow-in-the-dark toys, clock dials, etc.

counted by naked eye. The major breakthrough occurred when it was realized that scintillations, in turn, can induce electric signals in a well designed electrical arrangement.

The idea is simple. A primary radiation emits scintillations as it passes through the phosphor. The wavelengths of scintillations are characteristic of a phosphor. The intensity of scintillations depends on the type of radiation and their energies. We direct these scintillations to fall on a photo detector to produce pulses of electric current or voltage proportional to the energy deposits of the primary radiation. The response times of scintillators can be made to be sub-nano seconds to achieve very precise time relations among several types of radiations coming from a single decay or reaction event.

Thus, a scintillation detector response for a species of radiation may be quantified as follows:

- the number of scintillations emitted (N_s) is proportional to the energy deposited by the radiation (E)

- the number of photoelectrons (N_e) emitted by the photocathode is proportional to the number of scintillations (N_s)

- electric current or voltage from the detector (I or V) is proportional to the number of photoelectrons (N_e)

Thus the electric current or voltage (I or V) is proportional to the energy deposited (E).

$$E = k_V V \tag{9.2}$$

$$E = k_I I \tag{9.3}$$

depending on whether the detector measures current or voltage. k_V and k_I are constants for detector operating conditions such as applied voltages and amplifications we arrange to make the signals usable for succeeding stages or digital systems. For each measurement, we optimize the constants. We determine these constants from measurements of outputs for emissions from reference standards. It should be noted that these constants may vary with time, ambient temperature, etc. Thus one needs to calibrate detectors on a regular basis and adjust the settings to achieve the desired results.

9.6.1 Scintillation Materials

A wide variety of materials are available. They may be as simple as plastic scintillators or NaI (sodium iodide) or more recent crystals such as BGO ($Bi_4Ge_3O_{12}$), GSO (Gd_2SiO_5), etc. The choice of crystal depends on various factors:

(i) durability of the crystal in terms of mechanical strength, radiation hardness and other considerations

(ii) cost and availability of the material

(iii) the application at hand

Thallium doped sodium iodide [NaI(Tl)] crystals have been a workhorse for gamma energy measurements for about 70 years. The iodine ($Z = 53$) of high atomic number enhances the photoelectric effect, a phenomenon by which a photon deposits all its energy in a single interaction. A portion of the energy deposit is re-emitted as scintillations. Thallium is doped in the crystal to optimize the wavelengths of scintillations to match the sensitivity of the photo sensor (see below). While NaI(Tl) is still widely used, the quest for new scintillators is still on-going to overcome some limitations of NaI(Tl).

NaI(Tl) is hygroscopic, necessitating that it is hermetically sealed in metal to prevent hydration. The metal creates a dead layer and it can also be a source background due to secondary radiations. The energy resolution[17] of NaI(Tl) scintillators amounts to about 5–10%, a limitation for energy and intensity measurements of sources with multiple radiations.

Another consideration of a detector is the response time. Response time is a measure of how fast a detector sends a signal to the external

---

[17]The energy resolution of a detector is a measure of how close energies of two radiations can be and still be seen as two distinct radiations. It is expressed as

$$100 \times \frac{\Delta E}{E} \% \qquad (9.4)$$

where E is the energy at which two radiations separated by ΔE are seen as two distinct radiations. For example, 10 % resolution at 661 keV (^{137}Cs emission) gamma energy means that the two photons must be at least 66 keV apart to be seen as two radiations.

circuitry from the moment radiation is incident on it. Clearly, detectors with a faster response are better suited for accurate measurement of time correlations among multiple radiations from nuclear decays or particle reactions or in the environment of high intensity fluxes.

The following is a description of the terminology used to describe scintillation detector characteristics.

Table 9.1 lists properties of a few inorganic scintillators of interest. Among them, all crystals except sodium iodide are non-hygroscopic. But we pay the price that their relative outputs are lower than that of sodium iodide. The cost, radiation hardness, ease of use, etc. are other considerations in choosing materials for this purpose.

TABLE 9.1: Characteristics of a few inorganic scintillators.

Scintillator name	X_0	τ_{decay} (ns)	λ_{max} (nm)	Relative output
NaI(Tl)	2.59	245	410	100
BGO ($Bi_4Ge_3O_{12}$)	1.12	300	480	15
GSO ($Gd_2SiO_5(Ce)$)	1.38	50	430	25
LSO ($Lu_2SiO_5(Ce)$)	1.1	40	420	75
PbWO$_4$	0.89	30	450	1.3

Scintillator name The commercial name of the crystal and the chemical composition. NaI is doped with Tl and GSO and LSO are doped with Ce to optimize the response of those crystals as detectors.

X_0 is the radiation length, a parameter we introduced in interactions of photons. The intensity of photons drops to 36.8% as they travel one radiation length in the medium. Shorter radiation lengths afford compact crystal sizes.

τ_{decay} The decay time is a measure of dead time, the minimum time interval between two successive radiations to be detected in the same detector volume. Clearly, crystals of shorter decay times are better suited for high intensity measurements.

λ_{max} is the peak wavelength of scintillation emissions. The photo sensor should be matched to the range of wavelengths of the scintillator for optimal performance.

Relative output is a measure of the magnitude of the output signal. For historical reasons, all crystals are compared with the output of sodium iodide of the same size. For example, an LSO crystal generates about a 75% signal as of NaI of the same size, operating with a similar electrical arrangement, while GSO and BGO do not fare as well, with about 25% and 15% outputs. The lowest is PbWO4 at 1.3%.

For photons of much higher energies ($>$ several 10's of MeV's), where electromagnetic shower mechanisms dominate, scintillator slabs of $PbWO_4$ (lead tungstate) are used. As seen in Table 9.1, the light output is only 1.3% in comparison to NaI(Tl). However, this crystal is used for very high energy photons. The large energy deposits compensate for this deficiency, and short radiation lengths make the system very compact and affordable. Several high energy physics experiments make extensive use of this crystal.

9.7 Cherenkov Detectors

As we have seen in Chapter 5, relativistic charged particles moving in a medium of refractive index n at speeds greater than c/n emit Cherenkov light. These detectors have been extensively used in nuclear and particle physics experiments to distinguish between charged particles of different masses. More recently, they have found uses in synchrotron light sources to monitor the beam losses in storage rings and also for diagnostics of laser induced plasmas. The most extensive use is made in measurements where several different charged particles, all of the same momentum, are selected and the task is to separately identify each of them by their masses. We make use of simple kinematic relations connecting the mass, momentum and velocity of a particle and the characteristics of Cherenkov radiation. For a relativistic particle of speed $v = \beta c$ and momentum p, the mass m is given by

$$m = \frac{p\sqrt{1-\beta^2}}{\beta c} \tag{9.5}$$

A charged particle moving in a medium emits Cherenkov light of wavelength λ at an angle θ given by

$$\cos\theta = \frac{1}{n\beta} \tag{9.6}$$

The optical refractive index n of materials varies with wavelength and it is generally written as

$$n(\lambda) = B + \frac{C}{\lambda^2} + \frac{D}{\lambda^4} + \ldots \tag{9.7}$$

where B, C, D, etc. are positive numbers specific to each medium, which establishes that shorter wavelengths are of a higher index. The minimum wavelength or maximum energy of a Cherenkov emission is given by

$$n(\lambda_{min}) \times \beta = 1 \tag{9.8}$$

Thus particles of the same momentum but different masses are of different β values and emit light at angles specific to the mass of each particle. The Cherenkov light intensity is proportional to the distance a particle travels in the medium. A detailed calculation involving electrodynamics and quantum mechanics gives the intensity of photons (N_c) per unit length in the medium as

$$N_c = \frac{\alpha}{\hbar c} \times E_c \left[1 - \frac{1}{n^2\beta^2} \right] \quad \text{for } n\beta > 1 \tag{9.9}$$

$$\sim 370 \times E_c \left[1 - \frac{1}{n^2\beta^2} \right] \quad \text{per 1 cm path length} \tag{9.10}$$

where E_c is the energy of a Cherenkov photon in units of electron volts.

Example 9.1

Let us consider a particle beam composed of electrons, pions, kaons and protons moving in a medium of refractive index 1.3 at the Cherenkov light of 410 nm wavelength ($E_c = 3$ eV). All particles are of momentum $p = 1$ GeV/c.

Particle	Mass MeV/c^2	β for $p = 1$ GeV/c	$\theta = cos^{-1}(n\beta)$ degrees	N_c
Electron	0.511	0.9999	39.7	453
Pion	139.5	0.9904	39.0	440
Kaon	493.5	0.897	30.27	282
Proton	938.5	0.729	no radiation	

From the above, we can conclude that a particle is a proton if there is no Cherenkov radiation. The angular separation between an electron and a pion is not large and it would be difficult to separately identify them. Also, the Cherenkov light of an electron is the brightest.

Early Cherenkov detectors were used in threshold mode, i.e., the absence or presence of Cherenkov light is a test to distinguish particles. More recent versions make elaborate arrangements to measure the cone angle of emitted Cherenkov radiation by imaging the rings of light. In the above example, a cone with a 30 degree angle with the path of the particle as the axis identifies a kaon as the source of Cherenkov light.

9.8 Photosensors

Once we choose a scintillation medium, we need to couple it to a photon sensor to convert the optical signal to an electric one. A photon sensor functions on the physics of the photoelectric effect or photoconductivity. We encountered the photoelectric effect in our discussion of photon interactions. Invented in the 1930s, photo multiplier tubes (PMT) are the most common sensors based on the photoelectric effect. PMTs actually multiply the number of electrons. They are constructed with a cathode (called a photocathode) and an anode with several electrodes called dynodes. The photocathode, as its name implies, is held at a lower potential than other electrodes. Its surface is coated with a material which emits electrons when light is incident on it. The geometry and potentials are such that these photoelectrons gain energy on their way to the next dynode and they liberate more electrons from it. The electrons from the dynode migrate to the next dynode and more electrons are liberated. This multiplication continues until they are finally collected at the anode. This current is drawn out as a voltage pulse across an external resistor. If the assembly is made up of n dynodes, each providing a multiplication factor of m, then the overall gain is m^n. It is quite common that PMTs of 10–14 stages with a gain of about 10^6, for a gain of about $m = 4$ for each dynode. That is, on the average, for each electron incident on a dynode, four electrons will go to the next stage.

Very recently, new photosensors have become increasingly popular. They try to overcome some of the limitations of PMTs, such as having to work with high voltages of about 1–3 kV and not being functional in magnetic field environments. Increasingly, semiconductor technology[18] is contributing to this end. Currently, there are three serious contenders: avalanche photodiodes, pin diodes, and silicon photomultipliers. They are all based on a simple principle that current does not flow when diodes are reverse biased, i.e., when the anode is at a lower potential than the cathode. In steady state, there is no current. When a photon enters the diode, it liberates electrons, which are then accelerated to the anode. Because of the small gap of a few micrometers between the cathode and anode, a few volts results in a very high field gradient. As the electrons are accelerated they cause secondary ionizations, resulting in an avalanche of charge carriers. Hence the name avalanche photodiode. They are also called Geiger mode avalanche as the charge carriers saturate the system in this arrangement. Invented about 20 years ago, they are still developing and they will replace PMTs in all the medical imaging machines, such as PET scanners, some time in the near future.

In this chapter, we emphasize that radiation detectors rely on the interactions of radiations with matter. The choice of materials depends

---

[18]Semiconductors play increasingly important roles in photon sensors and also as detectors. A brief description of the physics can be found in the appendix.

on the ability to extract the interaction effects in optical or electrical signals. From there, it is the electronic arrangement that derives the physical information about radiation with analytical and computational tools.

10

Measurement Techniques

The purpose and scope of radiation measurements vary widely from one setting to another. It may be that we are engaged in measurements of

1. intensity levels in a known radiations environment to determine hazard levels

2. energies and/or intensities of various types of radiations in a complex environment

3. energy correlations of various types of radiations to build decay schemes of nuclei

4. energy and time correlations of various types of radiations to determine decay schemes and lifetimes.

5. identifying the species of radiations, their energies, energy and time correlations, etc.

The complexity of a radiation environment and the information we seek from a measurement are what dictate how elaborate an experimental setup should be and which detector elements are best suited, but the physics principles of measurements remain simple and straightforward. The physics observables are momentum, speed, energy deposits and ionization. All the information we deduce about nuclear radiation relies on some combination of these observables. It may be that a few alternate experimental arrangements afford the results one is after.

In earlier times, instrumentations were such that each piece of equipment was a big box to serve a specific purpose and we could identify its role uniquely in the arrangement. Thanks to advanced technologies, today's arrangements are quite compact and many of the hardware functions are delegated to software logic which can be programmed either during the measurement or in a leisurely analysis at a later time. In many instances, these developments make the measurements more versatile and more efficient. In some cases, we can do experiments which were inconceivable in the past. However, we should emphasize that the physics principles of measurement and the interpretations remain the same. Thus the emphasis of this chapter is to introduce those physics principles and measurement logic and illustrate them with some examples.

10.1 Intensity Measurements

FIGURE 10.1: A simple detection system. A is the detector, B is an amplifier and C is a scalar counter.

These are the simplest measurements, especially when we know the radiation environment. All we seek to measure is the number of photons, neutrons or other particles that pass through a fixed detector volume so that we can determine the background radiation levels. Quite often, we are interested in X-rays or gamma rays in a medical or natural environment. We may be concerned about neutron and gamma radiation levels in areas of a particle accelerator or a nuclear reactor setting. For these measurements, experimental arrangements consist of a simple setup shown as a block diagram in Figure 10.1.

A is a detector, which we should select to be suitable for the radiation to be detected. This choice depends on the type and probability of interactions of the radiation in the detector volume to generate a reliable response from the detector. We may use a simple Geiger–Müller (GM) counter for photons, while we should use a BF_3 or ^{3}He counter as a neutron detector. Some GM counters are built with a portion of the wall made of mica to be thin enough to let alpha and beta particles into detector volume. Such detectors can be used for measuring the levels of alpha, beta and gamma radiations. These detectors generally put out large pulses and we can count them in an electronic unit, called a scaler. A scaler consists of a counter which increases its output number by one each time the scaler receives an electric pulse. Thus there is a one-to-one correspondence between the output of a scaler and the number of particles interacting in the detector volume to generate a measurable electric pulse.

A hand-held radiation monitor is a compact device which contains a detector and a scaler as one unit. Its output is proportional to the number of photons interacting in the detector volume. An analog or digital display presents the data as counts per minute (CPM) or as radiation dose. It is important to note that GM counters measure CPM and conversion to dose depends on the energy of the radiations. Most GM counters showing dose readouts calibrate them with respect to either 662 keV photons from a ^{137}Cs source or 1.33 and 1.17 MeV photons from a ^{60}Co source. In general radiation level measurements, we can be sure of CPM data but dose readout is an approximate number. This is because general radiations consist of gamma rays of several different energies and each photon does not result in the same dose as ^{60}Co or ^{137}Cs gamma rays. As the general use of a hand-held device is to get a fairly accurate but not necessarily precise estimate of hazard levels, these units do an adequate job.

10.2   Energy Spectroscopy

These experiments aim to gather more information about radiation. We would like to know the species and energies of all radiations from a nuclear experiment or a radioactive source.

For charged particles, we may attempt different types of experiments. A first type is to measure the total kinetic energy of a particle by bringing it to rest in the active volume of a detector, and the result is an electric pulse proportional to the particle energy. The second type is to measure the momentum of a particle. In the first type, we measure energy deposits accurately, while the second one relies on deducing particle momenta by measuring trajectories of charged particles in known magnetic fields. As we discussed earlier, for a charged particle, with momentum p (GeV/c), its trajectory of radius of curvature R (meters) in a magnetic field of strength B (tesla) is given by

$$R = \frac{p}{0.3 \times n \times B} \qquad (10.1)$$

where $n = 1, 2, 3,...$ is the charge state of the particle.

For low speed particles, the information of kinetic energy (T) and momentum (p) are sufficient to determine the mass of the particle by the familiar Newtonian equation

$$p = \sqrt{2mT} \qquad (10.2)$$

For low speed particles, we can resort to energy loss measurements of charged particles. Here we exploit the feature that particles of the same momentum but of different masses have significantly different differential energy loss $[dE/dx]$ in a material medium.[1] But this involves a slightly more complex detector arrangement and we will discuss it later in this chapter.

For gamma ray spectroscopy, a sample can be placed in open air to carry out measurements. For charged particles of low energies, such as a few MeV alpha particles and betas from radioactive sources or at low energy accelerator experiments, the samples and detectors must be placed in a good vacuum system.

A Geiger–Müller counter cannot be used for energy measurements, as they are operated in a mode where the pulse amplitude is independent of energy deposits above the detection threshold. For energy measurements, it is essential to ensure that there is one to one correspondence between the energy deposit of radiation in the detector volume and the voltage amplitude or the total electric charge of the electric pulse output of the detector. This is called the linearity of response of the detector. We have to operate the detector and subsequent electronic units such as amplifiers to ensure that the final output which goes to a computer for sorting the pulses according to their amplitudes

---

[1] See Chapter 4.

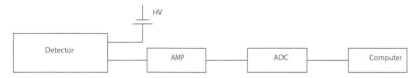

FIGURE 10.2: Typical spectroscopy setup with a detector, amplifier, analog to digital converter and computer for recording and processing data.

or the total charge will carry this linear correspondence. If the linearity is not maintained, at least there should still be one to one correspondence to the electrical output and the energy deposit, which may then be parametrized as a polynomial function.

Figure 10.2 is a sketch of the detector assembly for energy spectroscopy.

The first box is a transducer (detector) which converts the energy deposits to an electrical signal. The output is then fed to an amplifier which serves to shape the pulses both in form and amplitude to be compatible with a pulse sorter unit, usually called an analog to digital converter (ADC). The ADC output can be further processed by a computer.

The operating conditions of the detector, amplifier and ADC are adjusted to cover the dynamic range of interest and be able to determine the energies and intensities of radiations by simple calibration procedures.

The intensity data of gamma rays are not only of interest for scientific use but also for applications such as dosimetry. The radiation estimates from sources require the knowledge of energies of individual photons and also the gamma intensities since the dose imparted is a product of the energy of photons and the number emitted.

The efficiency of detection of a gamma ray depends on detector materials, operating conditions and also source-detector geometry. Several decades of works by nuclear researchers have yielded a vast amount of data for calibration purposes.

Table 10.1 lists a few commonly used isotopes for energy calibration. For convenience and reasons of economy, we would like to use isotopes of sufficiently long half-lives (preferably a few years) with a few gamma rays of well known energies and intensities.

TABLE 10.1: Energies and intensities of some gamma rays emitted by the most commonly used radioactive isotopes.

Isotope	Half-life (years)	Energy [keV]	Intensity [%]
^{57}Co	0.75	6.4	49
		14.4	9
		122.2	86
		136.5	11
^{133}Ba	10.5	81	33
		276.4	7
		302.8	18
		356	62
		383.8	9
^{137}Cs	30	661	85
^{22}Na	2.6	511	181[2]
		1274.5	100
^{60}Co	5.27	1173.2	100
		1332.5	100
^{152}Eu	13.5	121.8	29
		244.7	8
		444	3.1
		841.6	14
		963.4	12
		1085.9	10
		1112	14
		1408	21

Note: The energies are rounded off to 1/10 keV. The intensities are in percentage for 100 decays and accurate to 1%. In the case of positron emission decays, there are two 511 keV photons for each positron emission. Thus the maximum intensity possible for these decays is 200%. As discussed in Chapter 3, electron capture can offer an alternate decay path to positron emission. Thus the intensities of 511 keV may be less than 200%.

---

[2]The intensity of 181% for 511 keV indicates that 90.5% of ^{22}Na decays are by positron emission and the rest (9.5%) are by electron capture, both resulting in ^{22}Ne as the daughter product.

In addition, some naturally occurring isotopes provide good energy calibrations. For example, the isotope ^{40}K of 1.3 billion years half-life is commonly found in concrete walls. It decays by emitting a beta particle to ^{40}Ca, which subsequently emits a gamma ray of 1465 keV energy. In geographical regions with uranium deposits, we find ^{208}Tl, a decay product of natural 4n series with measurable intensities of 2614 and 1621 keV and a few more low energy gamma rays. The interesting point is that this naturally occurring radioactivity extends the energy calibration range to higher energies than is otherwise possible. Of course, at accelerators and reactors, we can produce short lived activities to this end.

10.3 Coincidence Measurements for Energy Spectroscopy

A look at the NNDC website for gamma cascades in decays of nuclear isotopes might leave one wondering how these complicated schemes are arrived at. While the data collection and analysis involve ingenuity and attention to detail, the logic is pretty straightforward. It is based on coincidence measurements, where two or more detectors are arranged to collect energy deposits of individual detector elements and also determine the relative time intervals among radiations arriving at the detectors.

The time information depends on the response of the detectors. Some phosphors, gas counters or semiconductors are of slow response times (several microseconds or longer). The electronics can be designed to collect the information at the initial stages of pulse formation to enhance the speed of response. Currently, these systems can respond at sub-nanosecond speeds.

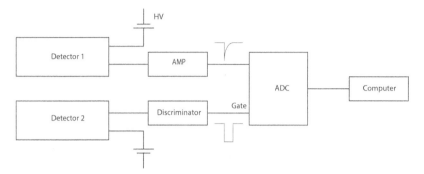

FIGURE 10.3: Two detector timing coincidence setup. The ADC is triggered to record an incoming signal from the amp only when a gate signal is sent out from the discriminator and only for a time window τ, usually a few hundred nanoseconds or a few microseconds.

10.3.1   Time Correlation Measurements

When several emissions occur in radioactive decays or nuclear or particle interactions, time correlations among the signals in a set of detectors involving coincidence techniques are valuable data. They provide extensive knowledge of the properties of radiations and those of the nuclear levels or particles of interest for scientific and technological purposes.

The basic idea is very simple. Say we examine the gamma spectrum of a ^{60}Co radioactive source. From the Q-value (see Chapter 3), we know that the beta decay of this isotope can populate nuclear levels of up to 2.8 MeV in ^{60}Ni. Thus, from the energy conservation principle, we can expect gamma rays of up to that energy. The gamma spectrum measured with a single detector exhibits 1.33 and 1.17 MeV gamma rays. From a single detector measurement, this is all the information we can obtain. We cannot decide whether they arise as cascades in the decays of single ^{60}Co isotopes or if they are uncorrelated emissions in which each ^{60}Co nucleus emits a single gamma ray.

A two detector arrangement, as seen in Figure 10.3, can be set up to let 1.33 MeV gamma ray be registered in one of them and 1.17 MeV gamma rays be recorded in the second detector. Let us set an electronic logic arrangement with a resolving time of τ seconds. Each time a detector sends a signal, the system will wait for up to τ seconds for the second detector's signal for the system to identify a coincidence event. If no signal from the second detector occurs, the event is lost. In modern systems, electronic time resolutions of about a few picoseconds (10^{-12} seconds) are possible but the detector response times are at least a few hundred picoseconds. In an experiment like this, we can determine that the two photons in ^{60}Co mainly arise from the decay of a 2.5 MeV level in two stages:

$$2.5 \xrightarrow{\text{1.17 MeV } \gamma} 1.33 \xrightarrow{\text{1.33 MeV } \gamma} 0 \text{ MeV}$$

excitations in an ^{60}Ni isotope. This measurement cannot determine if the lower level is at 1.33 or 1.17 MeV excitation. It is only from supplementary data from other nuclear physics experiments such as scattering of charged particles from a ^{60}Ni target that we can ascertain that the level occurs at 1.33 MeV.

One of the concerns is false coincidences, those which are not really from the same decay but appear to be so due to finite resolving times. We can understand the situation as follows.

Say the source strength and detector geometries are such that count rates in the detectors are N_1 and N_2 per second, respectively, for detectors 1 and 2. The resolving time is τ seconds. Each event in the detector activates the coincidence unit for τ seconds. Thus the first detector activates the system for τN_1 seconds for each second. As there are N_2 counts per second, the overlap time results in chance coincidence $\tau N_1 N_2$. In the same setting, detector 2 activates the coincidence unit for τN_2 seconds, which can overlap with detector 1 events for $\tau N_1 N_2$, amounting to chance coincidences (also known as accidental coincidences) of $2\tau N_1 N_2$.

The accidental coincidence in a two detector system is $2\tau N_1 N_2$.

In measurements, our interest is to keep the accidental coincidences to a minimum while not losing the genuine coincidences from single decay products, i.e., we have to minimize the product $2\tau N_1 N_2$. We cannot keep count rates too low as this will mean prolonged measurements. Based on the experimental situation, we have to optimize the count rates and resolving times.

Example 10.1

A 1 μC_i source of ^{60}Co will emit 3.7×10^4 pairs of 1.33 and 1.17 MeV energies. Assuming that they are fully intercepted by either detector 1 or 2, we have $N_1 = N_2 = 3.7 \times 10^4$/second.

The real coincidence rate is 3.7×10^4/second.

The accidental coincidence rate is

$$2\tau N_1 N_2 = 2\tau \times (3.7 \times 10^4)^2/\text{second}$$

The ratio of real to accidental rates is

$$= \frac{3.7 \times 10^4}{2 \times \tau \times 1.369 \times 10^9}$$

The higher this ratio is, the better the quality of the data. A good measurement requires that this ratio be at least one. Thus we require

$$\tau < \frac{10^{-4}}{7.4} \text{ seconds}$$

In this configuration, the resolving time should be less than 10 microseconds. Clearly, shorter resolving times give better results. In modern detector assemblies resolving times of a few nanoseconds are easily achieved.

10.3.2   Energy Correlation Measurements for Charged Particle Identification

In many experiments at particle accelerators for nuclear research, several species of particles are emitted. An important task entails the identification of each individual particle whether it is a proton, deuteron or something else. At low energies, the energy correlation measurements, exploiting the fact that the energy losses depend on the mass and charge of a particle (Bethe formula, Chapter 4), are employed. For this purpose, detector telescopes made of two or more detectors are used in coincidence conditions. In this arrangement, particles pass through thin detectors before coming to rest in the last thick detector. The kinetic energy of a particle is the sum of all energy deposits. In addition, we have the information of differential energy losses in the front detectors.

For a fixed energy deposit, the energy loss (dE/dx) is

$$\frac{dE}{dx} \propto mz^2 \tag{10.3}$$

For the same energy, the energy losses of $p, d, t, {}^3\mathrm{He}^{++}, \alpha$ particles are in the ratios

$$\frac{dE}{dx_p} : \frac{dE}{dx_d} : \frac{dE}{dx_t} : \frac{dE}{dx_{{}^3\mathrm{He}}} : \frac{dE}{dx_\alpha} = 1 : 2 : 3 : 12 : 16$$

Figure 10.4 illustrates the power of this technique. It is the result of an experiment in which 62 MeV energy protons were incident on a silicon target. The energy deposits were measured in a three detector telescope arrangement. The figure shows the plot of energy deposits in the front thin detector (ΔE_1) plotted against those in the second detector (ΔE_2). The first thing we note is that there are five distinct dark bands of events, each corresponding to one species of particle, labeled protons, deuterons, tritons, ${}^3\mathrm{He}$ and α. The bands of hydrogen isotopes are well separated from helium isotopes, reflecting the z^2 dependence of energy losses. All particles lose less and less energy in the front detector with increasing kinetic energies, in agreement with theoretical prediction.

10.4   Timing Spectroscopy

While we emphasized the time correlations for energy spectroscopy, time measurements are of interest for several other purposes. A few possible applications are

(i) time of flight measurements to measure the speeds of particles. This is an extremely useful technique for particle identification. As we mentioned before, a particle's momentum (p), kinetic energy (T) and mass (m) are related to speed (βc) by

$$p = m\beta\gamma c$$

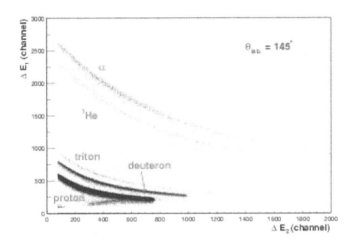

FIGURE 10.4: Energy loss plots for p, d, t, ^{3}He, α particles emitted as 62 MeV protons interact with Si nuclei. Bands of events for each species of particle are clearly seen. (Taken from C. Dufuquez et al., American Institute of Physics Conference Proceedings, Vol. 789, pages 941–944, 2005.)

and

$$T = (\gamma - 1)\,mc^2$$

Energy or momentum measurements supplemented with that of speed identify the particles uniquely.

(ii) Also, for massive neutrons, which do not leave ionization trails, the time of travel to the detector from a production point is a good measure of its energy. This technique is extensively used.

(iii) In a radiation environment of neutrons and gamma rays, these measurements can serve to distinguish between them.

(iv) Lifetime measurements of short lived nuclear levels or particles

In these measurements, one is interested in accurate time information beyond simple resolution concerns because they would reflect on the precision of physical observables.

A general layout of a two-detector timing measurement is shown in Figure 10.5. The discriminators generate fast logic signals. The time digitizer records the time intervals between the two detectors, to be processed by the computer.

10.4.1   Time of Flight Measurements

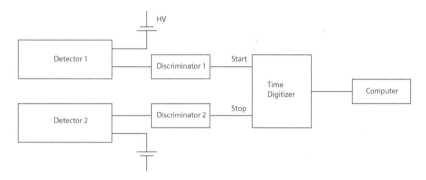

FIGURE 10.5: Two detector coincidence timing setup. The difference in time between the start signal and the stop signal is output as a voltage pulse recorded and interpreted by the computer.

In these measurements, a time marker for a particle leaving a point A is generated. Another detector located at a point B, separated by a known distance L from the point A registers the passage of the same particle through it. A time digitizer or a time to amplitude converter, which then is fed to a digitizer, uses one of the signals as a start time and the second one as the stop time. This time interval constitutes the time of flight of the particle from point A to B. It is easy to recognize that this technique is no different from timing the speed of an athlete in a swimming competition or running competition.

For a particle of mass m moving at a speed βc, we know the kinetic energy T is given by

$$T = (\gamma - 1)\,mc^2 = \left[\frac{1}{\sqrt{1 - \beta^2}} - 1\right] mc^2 \tag{10.4}$$

With speed $v = \beta c$, the time elapsed is

$$t = \frac{L}{\beta c} \tag{10.5}$$

In particle speeds, we refer to the time in units of nanoseconds or sub-nanoseconds. In these units,

$$c = 3 \times 10^8 \text{ m/s} = 0.3 \text{ m/ns}$$

We can write the expression for kinetic energy

$$T = \left[\frac{1}{\sqrt{1 - \left(\frac{L}{0.3t}\right)^2}} - 1 \right] mc^2 \tag{10.6}$$

Here t is measured in nanoseconds and L is measured in meters. If mc^2 is expressed in energy units (MeV), the kinetic energy is also in matching energy units (i.e., MeV).

Example 10.2

A proton traverses 1 m distance in 5 ns. What is its kinetic energy? Here $t = 5$, and the proton's rest energy is $mc^2 = 938.5$ MeV.

$$T = \left[\frac{1}{\sqrt{1 - \frac{1}{25 \times 0.09}}} - 1 \right] 938.5 \text{ MeV} = 320.5 \text{ MeV}$$

An alpha particle of rest energy $mc^2 = 3.727$ GeV traverses a 5 m distance in 20 ns. What is its kinetic energy?

$$T = \left[\frac{1}{\sqrt{1 - \frac{25}{400 \times 0.09}}} - 1 \right] 3.272 \text{ GeV} = 3.015 \text{ GeV}$$

As long as we express the time in units of nanoseconds and the length in meters, the kinetic energy is in the same units of rest energy.

If the time interval is measured as t with a precision δt as the particle passes through a distance L, the speed and the error in the measurement are given by

$$\frac{\delta\beta}{\beta} = \frac{\delta t}{t} \tag{10.7}$$

Also, by straightforward differentiation, we get the error in the kinetic energy (δT) due to the error in the speed measurement as follows:

$$\delta t = \frac{mc^2\beta}{\left(1-\beta^2\right)^{3/2}}\delta\beta = mc^2\beta\gamma^3\delta\beta \qquad (10.8)$$

The error in the speed measurement is due to errors in measurements of the distance and time.

With speed $\beta c = \frac{L}{t}$, we have

$$c\delta\beta = \frac{\delta L}{t} + \frac{L\delta t}{t^2} \qquad (10.9)$$

or the fractional error in speed measurement $\frac{\delta\beta}{\beta}$ is given by

$$\frac{\delta\beta}{\beta} = \frac{\delta L}{L} + \frac{\delta t}{t} \qquad (10.10)$$

In a measurement, we can keep the distance between the detectors constant and the fractional error due to the error in distance measurement is the same for neutrons of all energies. Below, we will neglect the δL dependence and focus on the error in time measurement. The fractional error due to the time measurement varies with neutron energies and it depends on the electronic equipment and resolving time of the detector arrangement.

We can write the fractional error in the kinetic energy measurement as

$$\frac{\delta T}{T} = \frac{mc^2\beta\gamma^3\delta\beta}{(\gamma-1)mc^2} = \frac{\beta\gamma^3}{(\gamma-1)}\frac{\beta\delta t}{t} = \frac{\beta^2\gamma^3}{(\gamma-1)}\frac{\beta c}{L}\delta t \qquad (10.11)$$

Example 10.3

In Example 10.2 protons are of kinetic energy 320.5 MeV and speed $v = \beta c = 0.66c$. The Lorentz factor is

$$\gamma = \frac{m+T}{m} = 1.34$$

In the same example, the alphas are of kinetic energy 3.015 GeV and speed $v = \beta c = 0.833c$. The Lorentz factor is

$$\gamma = \frac{m+T}{m} = 1.81$$

If we assign $\delta t = 0.1$ ns, which is about the best precision we can get today, we find the error in energy estimates to be

$$\frac{\delta T}{T} = \frac{0.3 \times (0.66 \times 1.34)^3}{1.34 - 1} \times \frac{0.1}{1} = 0.6 \text{ for protons} \qquad (10.12)$$

and

$$\frac{\delta T}{T} = \frac{0.3 \times (0.866 \times 1.81)^3}{1.34 - 1} \times \frac{0.1}{5} = 0.025 \text{ for alpha particles} \qquad (10.13)$$

We are able to determine the proton energies to about 6% precision and alphas to about 2.5% precision. The improved precision for alphas is achieved by increasing the flightpath by a factor of 4, as the speeds and Lorentz factors of the protons and alphas, in this example, are of comparable magnitudes.

From the above equation, we note that we can minimize the fractional error in kinetic energies by

- reducing the error in time measurements

- increasing the flight distance for the same error in time measurement

Designers weigh the need, economical, technical and spatial considerations in attempts to get the best possible result.

10.5 Particle Identification by Time of Flight

In many experimental arrangements, several charged particles of different species are emitted. It is the task of an experimenter to identify them individually for the physics problem at hand. As mentioned earlier, one arrangement consists of measuring the momentum of each particle and the next step is to measure their speed. In most common arrangements, the times of flight of particles of known momentum, say p, are measured. We know that

$$p = mc\gamma\beta = \frac{mc^2\gamma\beta}{c}$$

If the momentum of particles and flightpaths are fixed, then time measurements determine the $\beta\gamma$ of each particle and the masses are determined.

Example 10.4

In Example 10.2 the protons have a momentum of $p = 938.5 \times 0.66 \times 1.34 = 830$ MeV/c and a flight time of 5 ns/m.

For each particle, we can calculate the β from the following.

$$\beta = \frac{p}{E} = \frac{830}{\left(830^2 + (mc^2)^2\right)^{\frac{1}{2}}}$$

and the time of flight $= \frac{1}{\beta c} = \frac{3.33}{\beta}$ ns.

Table 10.2 lists the flight times of a few species of particles of momentum 830 MeV/c and their flight times per meter in the same measurement.

From the table, we see that the time difference per unit path length for the light charged particles (electrons and pions) is too small to be distinguishable. Long path lengths of several meters might accomplish the separation. For heavier particles, the situation is excellent. Even worse time resolutions of about 1 ns or shorter path lengths of several centimeters will allow us to distinguish among these particles.

One point worth noting is that the electrons and pions are of a speed very close to that of light. For these particles, it is almost inconceivable that this technique will work to identify them.

TABLE 10.2: Flight times for various particles with momentum $p = 830$ MeV/c.

Particle	Mass (MeV/c^2)	β for $p = 830$ MeV/c	Flight time 1 m $= \frac{3.3}{\beta}$ [ns]
Electron	0.511	0.999	3.3
Pion	139.5	0.986	3.35
Kaon	493	0.86	3.87
Proton	938	0.66	5.0
Deuteron	1876	0.404	8.24
Triton	2809	0.28	11.9
Alpha	3727	0.217	15.3

For increasing momenta and lighter masses, the resolution gets worse. We should resort to other techniques.

10.5.1 Particle Discrimination by Cherenkov Radiation Emission

As we mentioned in the Section 5.9, Cherenkov radiation detection provides a powerful tool at higher energies and it is extensively used. It relies on the detection of light emission.

Thus, while the detector materials are different from that of conventional scintillators, the electric pulse generation relies on the tools commonly used in photon detectors, such as photomultiplier tubes, avalanche diodes, etc. Also, we are interested only in knowing if the Cherenkov light is emitted by the passage of a particle in the detector volume and not so much in the quantitative measure. This makes life simpler as we can design a fast response system to generate a logic signal for coincidence measurements.

The differential measurements of Cherenkov light are needed to identify the charged particle by determining the angle of emission. Quite often, a simpler threshold measurement is used to decrease the data volume by avoiding undesirable triggers to the electronics and computers. We note that the lowest refractive index (n_{min}) below which a particle does not emit Cherenkov radiation is given by $n_{min} > \frac{1}{\beta}$.

Imagine a measurement at a high energy accelerator such as an electron or photon beam facility of several hundred MeV or a high energy accelerator such as those at Brookhaven, CERN, J-PARC, etc. Quite often, very high energy photons are emitted which, in turn, emit electrons and positrons due to the conversion processes. These showers propagate long ways in the target materials and detectors, generating signals. If we let these signals trigger all electronics and keep the computer systems busy, it will be almost impossible to record the reaction events of interest.

A simple arrangement is as follows. In a measurement to detect pions, kaons and other heavy charged particles of momenta mentioned above, we look for a detector material of optimum refractive index.

Table 10.3 lists the n_{min} for the emission of Cherenkov radiation by particles referred to in Example 10.4.

TABLE 10.3: Indices of refraction for various particles with momentum $p = 830$ MeV/c.

Particle	Mass (MeV/c^2)	β for $p = 830$ MeV/c	n_{min} for Cherenkov emission
Electron	0.511	0.999	1.001
Pion	139.5	0.986	1.014
Kaon	493	0.86	1.163
Proton	938	0.66	1.515
Deuteron	1876	0.404	2.475
Triton	2809	0.28	3.57
Alpha	3727	0.217	4.61

First, we recognize that the electrons emit Cherenkov radiation in practically any liquid or solid materials of refractive index greater than 1.01. We can arrange a gas Cherenkov counter (detector 1) with the pressure of the gas adjusted to correspond to a refractive index just about 1.01. Only electrons will generate Cherenkov radiation in this detector.

In recent years, aerogel counters, which are silicon materials into which air bubbles are introduced in a controlled way, have become

TABLE 10.4: Cherenkov logic for charged particles of momentum $p = 830$ MeV/c in three detectors of refractive indices $n = 1.01, 1.1$ and 1.3.

Detector 1 $n = 1.01$	Detector 2 $n = 1.1$	Detector 3 $n = 1.3$	Particle identification
yes	yes	yes	Electron
no	yes	yes	Pion
no	no	yes	Kaon
no	no	no	Proton or heavier particle

available. They can be made to have refractive indices as low as $n = 1.008$ and up to about 1.1. If we employ such a detector (detector 2), electrons and pions will emit the radiation and not the heavier particles.

We may consider a third detector (detector 3) of refractive index $n = 1.3$ (say a water medium). In this detector, electrons, pions and kaons emit radiation, but not protons and heavier particles.

The following logic table (Table 10.4) can be constructed with "yes" or "no" answers for the emission of Cherenkov radiation with a set of three detectors.

As we saw above, it is not difficult to separately identify kaons, protons and heavier particles from time of flight information. We may then choose to work with just two detectors and time of flight information for economy and less complexity.

From Table 10.4, we see that it is only electrons which will give a "yes" signal in detector 1. For those situations where we do not want signals of electron arrival to keep the electronics and computers busy, we can use this signal to send a "veto" signal to the subsequent system. This will prevent the electronic signals from propagating to the next level of data manangement and discard the event. This allows the system to get ready for another particle once the detector recovers. Usually, this recovery occurs within a few nanoseconds time, whereas a computer system could have taken a few milliseconds or longer, depending on the complexity of the event.

Thus Cherenkov detectors have extensive uses in particle identification and also in the reduction of data volume. Furthermore, they help to pick up signals of lower frequently appearing hadrons (pions, kaons, protons, etc.) in the presence of a high background of electrons. These data could not be obtained without Cherenkov detectors or other sophisticated arrangements.

11

Nuclear Techniques — A Few Applications

11.1 Introduction

Nuclear radiations and their measurements are used in basic sciences, industrial, agricultural, energy and medical applications.

11.1.1 Some Applications in Agriculture

Most of the agricultural applications are in the sterilization of food and mutations of plants and seeds. This is very commonly done with beams from electron accelerators or gamma radiations from sources such as ^{60}Co or ^{137}Cs. The sterilization of food by heating and exposure to solar radiation is a centuries old technique. Drying fruits and vegetables by exposure to the sun's radiation (irradiating) has been extensively used in developing countries. Sterilization by ultraviolet and microwaves is also very commonly employed. It is important to remember that gamma rays belong to the same family, viz., electromagnetic radiation, as ultraviolet and infrared radiation. Usually food irradiation is carried out by letting it pass by the source on a conveyor belt where it is exposed to radiation from a gamma emitting source or bremsstrahlung emitted by electrons from an accelerator.

The high energy gamma radiation penetrates deeper into materials and induces processes not possible at low energies. For example, Compton scattering and the photoelectric effect are important phenomena, dislodging electrons and rearranging chemical structures. Energies of a few MeV are not high enough to induce nuclear reactions ejecting neutrons, protons or other nuclei. From the binding energy and separation energy estimates (Chapter 3), we know the energy required to cause artificial transmutation which may induce radioactivity. For most stable elements, the separation energies are several MeV. For the most commonly used irradiation sources, the energies of gamma rays (1.17, 1.33 MeV for ^{60}Co and 0.661 MeV for ^{137}Cs) are too low to cause concern of induced radioactivity.

For agriculture and food applications, three terms are used to classify doses. As we know, a Gray is the deposit of one Joule energy per kilogram of matter.

The terminology and effects of these doses on food, plants and pathogens are as follows:

Radurization Low doses less than 1 kGy. Low doses are used to slow down ripening of fruits or vegetables to increase shelf life and to control insects in grain.

Radicidation Medium doses 1–10 kGy. Medium doses are sufficient to kill bacteria in meats and to prevent mold growth on fruit.

Rapperitization High doses greater than 10 kGy. High doses destroy all insects, bacteria, etc.

It is important to recognize that the emphasis is on the dose deposits rather than the energy of radiation. As we have seen in Chapter 5, the photoelectric effect is the only process in which photons deposit all their energy in the medium. In all other interactions, the energy is lost to the surrounding medium.

It is of interest to examine the relative merits of facilities of radioactive sources versus those of electron beams.

With a radioactive source such as ^{60}Co, which emits two gamma rays with the average energy of 1.25 MeV, the total deposit per decay is 2.5 MeV, assuming that photons lose all their energy in the target.

If the target is of 1 kg weight, each decay amounts to a dose of

$$\frac{\text{dose}}{\text{decay}} = 2.5 \times 1.6 \times 10^{-13} = 4 \times 10^{-13} \text{ Gy}$$

If we assume that gamma rays from all the decays of ^{60}Co are intercepted by the 1 kg target material, the 1 kGy dose deposition in 1 kg of matter requires

$$\text{Total number of decays} = \frac{10^3}{4 \times 10^{-13}} \text{ decays}$$
$$= 2.5 \times 10^{15} \text{ decays}$$

Total number of decays = source activity (A) × exposure (t)

or

$$At = 2.5 \times 10^{15} \text{ B} \cdot \text{s}$$
$$= 0.68 \times 10^5 \text{ Ci} \cdot \text{s}$$
$$= 68 \text{ kCi} \cdot \text{s}$$

This is an optimistic estimate for the following reasons.

• It is almost impossible to fully enclose the target material in the source. This is especially true for those operations in which we have to change targets very often. In most configurations, the source is set at a fixed point and the targets move on a conveyor belt without coming into contact with the source itself.

• We know that, unlike the behavior of charged particles, gamma rays can pass through materials without interacting in them. All agricultural materials are organic substances composed mainly of light elements such as carbon, nitrogen, etc. For these elements, the photoelectric effect is not the dominant mechanism. Thus, in practical situations, one uses radioactive sources of a few hundred or thousands of Curies.

Beams from electron sources have the advantage that they are directional. We can send electron beams of finite sizes along a straight line. As we have seen earlier, we can produce a narrow cone of a bremsstrahlung beam of photons to irradiate the materials. This is an advantage over radioactive sources, which emit gamma rays isotropically in all directions.

However, the disadvantage of electron beam facilities is that bremsstrahlung is not a monoenergetic beam but a continuous spectrum of intensity

$$I(E_\gamma) \propto \frac{1}{E_\gamma} \text{ for } E_\gamma < E_e \tag{11.1}$$

If the proportionality constant is C, then

$$I(E_\gamma) = \frac{C}{E_\gamma} \tag{11.2}$$

The contribution of photons of energy E_γ to the dose is

$$\text{Energy of gamma ray} \times \text{intensity} = E_\gamma \times \frac{C}{E_\gamma}$$

$$= C, \text{ independent of gamma energies}$$

If we consider that there are optimum energies for irradiation, the power in the remainder of the photon spectrum is wasted. At times, it can result in adverse effects.

For agriculture applications, one employs electron accelerators with beam energies of 10 MeV or less. Let us assume that we are working with a 10 MeV accelerator. If we assume that photons below 1.3 MeV energy contribute to the dose in the same manner as a ^{60}Co gamma ray, nearly 13% of the power of the output beam from the accelerator is used. Perhaps this estimate is a bit on the low side. But the main concern is that nearly 40% of the power is above 6 MeV photons and 20% of the power is 8 MeV photons. Photons above 8 MeV energy are prone to induce radioactivity through a (γ, n) reaction and also create a radiation background of neutrons.

From a physics perspective, we are simply concerned about energy deposits due to different interaction processes. However, it is not clear if the biological effects do not strongly depend on the energy deposits. If this is true, then bremsstrahlung may result in physiological effects in plants, fruits, etc. which are different from that of well defined photon energy interactions from radioactive sources.

On the practical side, irradiation by a radioactive source has the advantage that we can rely on the sources. They decay in a predictable manner and the source is always available. This may also be a disadvantage since we cannot switch off the radiation.

Electron beam accelerators, on the other hand, can be turned off when not in use. However, the reliability of the system and electric power consumption needs have to be addressed.

11.2 Applications in Medicine

Nuclear radiations have been playing increasingly important roles in medicine for both therapeutic and diagnostic purposes. Not only for human health, but animal care organizations also have been taking advantage of these tools. It may be amusing but interesting that radon therapy is practiced as alternative medicine for pain relief in arthritis, asthma, bronchitis, etc. People seeking a cure for their ailments travel to mine areas where the radon levels are about 1000 pCi/L of air, a factor of 200 or so higher than what environmental protection agencies consider to be maximum permissible levels.[1]

11.2.1 Radiological Imaging

We may trace the birth of radiological imaging to the well-known picture of the palm of Mrs. Roentgen showing her finger bones and wedding ring. The principle of X-ray imaging is different from that of conventional photography only in that the latter is based on the reflection of light from surfaces while X-ray imaging relies on the transmission and absorption of radiation. As we have seen earlier, the attenuation coefficient of X-rays varies with the X-ray energy and the material medium. As a general rule, higher energy X-rays are absorbed less than those of lower energies. Also, high Z material media absorb X-rays more than low Z materials. In a human body, the soft tissue is nearly transparent to the X-rays, while dense bones are more absorbing. That is how Roentgen was successful in producing dark images of his wife's fingers and her wedding ring.

11.2.1.1 Computed Tomography

Just as conventional photography has gone digital, so has X-ray imaging. A major development in X-ray imaging is computed tomography (CT). Tomography is a generic term derived from Greek. It generates figures or pictures of several slices of the solid object to gain detailed

---

[1] See Barbara Erickson, *The Therapeutic Use of Radon: A Biomedical Treatment in Europe; An Alternative Remedy in the United States,* in Dose Response, vol. 5, pages 48–62 (2007).

cross sectional views of it. It is very similar to an engineering drawing where we are presented with a top view, bottom view, side views or a cross sectional view, etc., to gain a full understanding of the object we build. Similarly, in a medical diagnostic process or industrial non-destructive testing, we want to know where a defect or tumor is, how wide and deep spread it is. We would also want to know where healthy organs are with respect to tumorous cells, so that we can avoid damaging the healthy ones.

Each image is labeled with the relative orientation of the radiation source, its beam profiles, object location and its geometrical details and that of the detectors and their performances. In former times, it would have meant that we move the source, the object and the detectors to several positions, which would be both time consuming and a painful process for patients and technicians. These procedures also introduce several errors due to time dependency and positioning systematic variations. The revolutionary trend of computers and instruments has made it possible to gather this information very easily and in much more detail and precision than in former times.

Simply stated, the computed tomography scan is the procedure where images are generated for predetermined relative orientations of the radiation beam, the object and the detector arrays.

Say there are a number of detectors $d_i, i = 1, 2, ... n$, each positioned at coordinates $r_i, i = 1, 2, ... n$.

The patient's organ to be imaged is located at the coordinates $r_p, p = 1, 2, ... m$.

The source illuminates the patient's organ uniformly and the transmitted radiation is recorded in the detectors d_i.

The intensity profiles recorded by the detectors constitute the data. It is then a matter of applying the principle of rectilinear propagation of radiation between the patient's body to individual detectors to reconstruct the images.

The data can be analyzed by a radiologist for various orientations of the patient in greater or lesser detail as needs arise, all with a software program command, without having to recall the patient for further imaging procedures. The information gathered is perhaps as good as a million times exposure with a single detector and a beam. Tomographic scans are thus very powerful tools for diagnostics either for medical purposes or industrial applications. The principles and concepts are the same while the end goals are different.

11.2.1.2 Single Photon Emission Computed Tomography (SPECT) and Positron Emission Tomography (PET)

X-ray imaging and the CT scan are similar to conventional photography with an external source of radiation. In astronomy or our human eye, images are formed by measuring the intensity profiles of radiation emitted by a source. In the same fashion, the SPECT and PET measure the radiation distribution from sources.

SPECT, as the name implies, employs radioactive isotopes which emit a single gamma ray in the radioactive decay of an atomic nucleus. The intensity distributions of gamma rays from individual nuclei are measured and analyzed to deduce the source distribution in the body of a patient. This is analogous to imaging a moon or star or a flame by detecting their emissions. The choice of isotopes is based on many considerations. A few criteria are listed below.

(i) The isotope must be sufficiently long lived but it should not be of too long a half-life. It must be of a few hours lifetime to allow for detailed measurements. If it is of too short a lifetime, then we have to introduce very high doses in the body, which is not desirable.

(ii) The yield of gamma rays must be high. If the branchings are small, we have to administer high doses. Also, it is preferred that other undesirable emissions be as few as possible. This is to minimize the radiation dose to patients.

(iii) We should be able to perform radiochemistry to prepare the compounds that the organ will accept.

(iv) It should be possible to make these isotopes in an affordable way.

(v) It should leave very little, preferably none, as residues in the body.

The quest for new isotopes is an on-going study. We list a few widely used isotopes and their properties as it pertains to radiations and imaging.

TABLE 11.1: A partial list of isotopes commonly used for SPECT imaging, half-lives, gamma energies and common applications.

Isotope	Half-life	Method of Preparation	SPECT gamma ray energy (keV)	Some Imaging Applications
^{99m}Tc	6.01 hours	Nuclear fission[2]	140	Diagnostics of coronary artery disease, studies of bone metabolism
^{123}I	133.3 hours	^{123}Te(p,n); ^{124}Te$(p,2n)$	159	Diagnostics of thyroid diseases
^{75}Se	120 days	^{75}As(p,n); ^{75}Se(γ,n)	74.9	Neuroendocrine tumors
^{67}Ga	61.8 hours	^{67}Zn(p,n); ^{68}Zn$(p,2n)$; ^{70}Zn(p,α)	93.3, 184.6, 300.2	Lymphoma, Infection
^{111}In	2.8 days	^{109}Ag$(\alpha,2n)$; ^{111}Cd(p,n)	171.3, 245.3	Infection, prostate cancer
^{133}Xe	5.24 days	^{130}Te(α,n)	81	Lung, global cerebral blood flow
^{201}Tl	3.01 days	^{202}Hg$(d,3n)$; ^{203}Tl(p,t)	167.4	Myocardial perfusion studies, recurrent brain tumor

From the table we notice that SPECT photons are generally about 100–300 keV. They are easily attenuated by dense materials such as bones. The images thus have good contrast between soft tissue and bone imaging.

Very similar to the CT scan, an array of detectors is arranged to collect the intensity distributions. A common configuration comprises a large scintillation detector (e.g., NaI(Tl)) crystal to which a few

---

[2]This isotope is in great demand for several medical applications. Other methods of production by, say, ^{98}Mo (n,γ), ^{100}Mo $(p,2n)$ and ^{100}Mo (γ,n), are being explored.

photomultipliers are coupled to derive the signals. Each photomultiplier, in general, is sensitive to the scintillations in the portion of the crystals that it overlaps. Long narrow lead collimators are coupled to the crystal surface so that only gamma rays emitted along the direction of collimator holes will generate signals. It is then assumed that the line joining the center of the photomultiplier receiving the signal to the patient is the path of the photon. The data taking is carried out for specific relative orientations of the crystals with respect to the patient. This data is fed into computer programs which reconstruct the source distribution in the patient's body and constitute the image. This will reflect the relative concentration of the radioactive isotope in the organ under study. The choice of isotope and its chemical composition are dictated by the physiological characteristics of the organ and its affinity to the isotope.

Limitations

SPECT is single photon emission counting. Thus it can only provide a line of sight of the photon direction but not where it exactly originates. However, the knowledge of physiological conditions and the fact that it is from the patient allow us to reconstruct the image fairly accurately. As photons undergo Compton scattering without much loss of energy and still change the direction of propagation, this constitutes an undesirable artifact. There are several proposals to correct for this effect by employing dual energy SPECT, which relies on the fact that the attenuation coefficients vary with the energy of the photon. One thus hopes to correct for the aberrations due to scattering or absorption artifacts. The work continues.

Positron Emission Tomography (PET)

This technique takes advantage of the fact that a positron (antiparticle of an electron) annihilates with an electron in the surroundings. As a consequence, two photons, each of 511 keV, moving back to back at an angle of 180 degrees are emitted. If we can set a two detector coincidence to register the two photons from a single decay, the line joining the two detectors gives the path of the photons, and the decay point of the radioactive isotope lies on that line, with a small correction for the migration of a positron from its birth point. If we can measure the difference in arrival times of the two photons at the detectors, we can also locate the positron annihilation point rather accurately. The task of imaging is to measure the relative distributions of the two 511 keV photon coincidences and then reconstruct the image. We can speed up the data taking procedure by using an array of detectors around the patient and measuring two photon coincidences. The data is then processed by algorithms to reconstruct the images just as in CT and SPECT.

Here again, the Compton scattering of photons can produce aberrations. If one of the photons is scattered, then it enters a detector different from the one it would have hit otherwise. If we simply rely on two detectors with signals, our reconstruction will be off. Another artifact is that one of the photons is absorbed. In this case, we lose information. Also, photons of 511 keV are quite transparent to soft tissue and they are not good for imaging it. Their use is in imaging bones and dense materials.

As we mentioned above, there have been several suggestions to create images with two photons of different energies, known as the dual energy imaging technique. This is intended to allow for in situ corrections of aberrations since the attenuation coefficients are strongly energy dependent, especially for photons of PET and SPECT energies.

The ^{18}F isotope of a half-life of 108 minutes is the most commonly used PET isotope. Because of its short half-life, it is locally produced near hospitals. The main production channel for the isotope is the $^{18}O(p,n)^{18}F$ reaction, which has a maximum cross section for about 8 MeV protons (see Figure 11.1). The fact that we need low energy proton beams for maximum yields makes a good market for not so expensive compact cyclotron facilities. Such facilities are found in several countries around the world.

Example 11.1

An 8 MeV proton beam facility employs a cylinder filled with water of an ^{18}O isotope of composition 2 atoms of hydrogen and 1 atom of ^{18}O-isotope for a molecular weight of 20.

The target has a cross sectional area of 1 cm^2 (radius = 0.56 cm) and is 1 cm long.

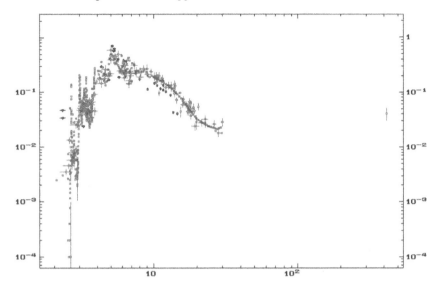

FIGURE 11.1: $^{18}O(p,n)^{18}F$ cross section as a function of proton energy. A broad maximum at about 8 MeV proton energy makes this channel a good choice.

The cross section of the $^{18}O(p,n)^{18}F$ reaction for 8 MeV protons is

$$0.3 \text{ barns} = 0.3 \times 10^{-24} \text{ cm}^2$$

The atom density of the target is

$$\frac{6.03 \times 10^{23}}{20} = 3 \times 10^{22}$$

for a target of 1 g/cm^2 thickness.

$$1 \ \mu\text{A proton beam} = \frac{10^{19} \times 10^{-6}}{1.6} = 6.25 \times 10^{12} \ \frac{\text{protons}}{\text{s}}$$

Constant rate of production of ^{18}F atoms:

$R = $ cross section $\times$ number of target atoms $\times$ number of protons/s

$$= 0.3 \times 10^{-24} \times 3 \times 10^{22} \times 6.25 \times 10^{12}$$

$$= 5.625 \times 10^{10}$$

At time t during irradiation with protons ($t < t_{irradiation}$), the number of atoms is given by

$$N(t) = \frac{R}{\lambda}\left(1 - e^{-\lambda t}\right)$$
$$= 5.625 \times 10^{10} \times \tau \times \left(1 - e^{-t/\tau}\right)$$
$$= 5.26 \times 10^{14} \times \left(1 - e^{-2.227 \times 10^{-4} \times t}\right)$$

The activity, A, at time t is

$$A(t) = \lambda N(t)$$
$$= R\left(1 - e^{-\lambda t}\right)$$
$$= 5.625 \times 10^{10} \times \left(1 - e^{-2.227 \times 10^{-4} \times t}\right) \text{ Bq}$$

The maximum activity for this production protocol of beam energy, current, and target combination (i.e., 1 μA current of 8 MeV protons on 1 g/cm^2 water of ^{18}O atoms) is 56.25 GBq = 1.52 Ci.

As we noted earlier, this maximum activity is reached for irradiation times longer than three mean-lives of the isotope, which is about 8 hours of irradiation. Needless to say, the operation protocols of irradiation times, target thickness and beam currents will aim to optimize the production rates.

For medical uses, radiochemistry is done to substitute an ^{18}F atom in a glucose chain to produce flourodeoxyglucose, known by its acronym FDG. Today's commercial medical cyclotrons put out proton beam currents of almost 1 A, while a few hundred milliamperes is quite common.

^{11}C ($t_{1/2} = 20.4$ m), ^{13}N ($t_{1/2} = 10$ m), and ^{15}O ($t_{1/2} = 2$ m) are three other commonly used isotopes. They are readily produced at cyclotron facilities. The very short half-lives of a few minutes for these isotopes dictate that they be produced near the diagnostic facilities.

In summary, CT, SPECT and PET employ very similar imaging modalities. CT relies on an external source of radiation. It leaves no residual radiation in the patient or in surroundings. In SPECT and PET, the bodies of patients are the sources of radiation administered to them.

11.3 Radiation Therapy

The aim of radiation therapy is to destroy cancerous cells in the body while doing as little damage, preferably no damage, to healthy organs. It is perhaps a miracle and certainly a blessing that tumorous cells are more readily killed while healthier cells are more resistant. Based on this knowledge, the therapy principles are very simple. Beams of radiation are directed into the tumor volume to deposit enough energy to irreversibly alter the physical/chemical composition of the tumor cell to result in its death. To this end, various species of radiations are employed.

11.3.1 Gamma Beams

Gamma beams were among the first tools for cancer treatment and they continue to be employed even today. As we discussed in detail, photon interactions with matter result in the exponential decay of the intensity of radiation. This means, while we can design radiation treatment protocols to maximize the dose at the tumor cells, there is a finite dose delivered at both deeper and surface layers of the patient. The task of radiation therapy is to optimize the dose protocols to minimize the damage to healthy cells.

11.3.2 Treatment with Radioactive Isotopes

The gamma radiation from ^{60}Co has been widely used to this end. It has a half-life of 5 years and emits hard gamma rays (1.17 and 1.33 MeV) which penetrate deep into the body. It is inexpensive to produce ^{60}Co at a nuclear reactor by the capture of neutrons by a ^{59}Co nucleus.

It is of interest to calculate the energy deposits of ^{60}Co photons in a human body. For illustrative purposes, we may assume the body to be composed mostly of water and to be of density $\rho = 1$ g/cm^3. We can ignore the contributions of other elements. The attenuation coefficients for ^{60}Co photons and that of average emission energy of 1.25 MeV are listed in Table 11.2. Since both 1.17 and 1.33 MeV photons are of the same intensity, the average values will serve the purpose.

TABLE 11.2: Attenuation coefficients of gamma rays in a water medium calculated from the XCOM website.

Photon Energy [MeV]	Attenuation Coefficient μ_ρ [cm^2/g] $= \mu$ [cm^{-1}] since $\rho = 1$ g/cm^3
15	0.0194
6.0	0.045
4.0	0.0581
1.33	0.1094
1.25	0.1129
1.17	0.1168

Note: Listed are the values of ^{60}Co emissions and 4, 6 and 15 MeV photons used by medical linear accelerators. Also listed is the coefficient of 1.25, the average photon energy of ^{60}Co photons. As the water medium is of density $\rho = 1$ g/cm^3, the mass attenuation coefficient (μ_ρ) and linear attenuation coefficient (μ) are numerically the same for this medium.

We can calculate the intensity of gamma rays in the human body from the equation

$$I = I_0 e^{-\mu x} \qquad (11.3)$$

where x is the length of the path of interaction in centimeters.

Also, the intensity (power) loss per unit length in the body is given by

$$dI = -\mu \times I_0 e^{-\mu x} = -0.1129 \cdot I_0 \cdot e^{-0.1129 \cdot x} \qquad (11.4)$$

If we assume a 30 cm width of a person, there is about 3% of the photon beam emerging from the other side of the person. Also, the intensity loss or energy deposits decrease slowly, being about 10%/cm on the source side and about 3%/cm on the farther side. Thus a tumor of 1 cm thickness at a depth of 10 cm in the body will receive only 4% of the power deposited in it, while the rest of the radiation dose is spread. We should remember that this is a ballpark figure. The presence of bones and other hard materials might improve the situation.

11.3.3   Gamma Knife

This is an improvement over the single beam radioactive source.

In this arrangement, multiple beams of a radioactive source channelled through a set of collimators are directed to the target volume. Figure 11.2 shows a five-beam arrangement of the radioactive source aiming the radiation at the target volume through the collimators. The collimators are tapered in a conical shape to minimize damage to the surrounding parts of the body. In a one beam arrangement, each volume would receive five times as much dose as in the multiple beam system. The advantages are clear.

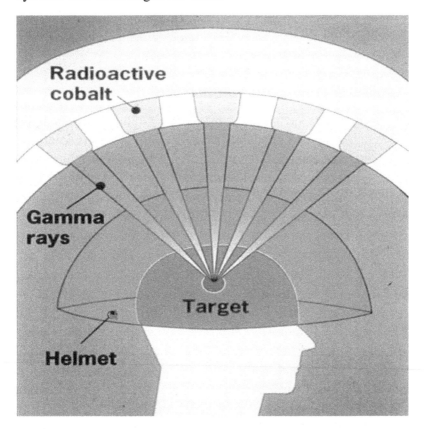

FIGURE 11.2: A schematic figure of a gamma ray knife. The cones are collimators which direct the photons to the target. They are designed such that beams from all collimators at the target. (Source: WikiCommons, Nuclear Regulatory Commission.)

11.3.4   Electron Beam Facilities

We have seen that a few MeV electrons readily emit radiations. Electron linear accelerators of 4–15 MeV are commonly used for these purposes. They are bremsstrahlung photon beam facilities and the statements with regard to agricultural applications of the relative merits of radioactive sources versus electron accelerators apply here. One reason for the desire for high energy beams is their penetrability. Table 11.2 lists the attenuation coefficients for 4, 6 and 15 MeV energies, which are less than half of that for ^{60}Co photons. It is easy to calculate that more than 40% of the beams of 15 MeV will emerge on the other side of a body of 30 cm thickness. While this observation may be somewhat disconcerting, we also note that more than 90% of photons are of energies less than 10 MeV.

11.3.5   Radiation Therapy with Particle Beams

Though arrangements such as the gamma knife minimize the radiation dose to surrounding volumes, they cannot eliminate it completely. It has been recognized that charged particles offer an advantage in this regard. We recall two features. The first is that charged particles travel finite distances in material media and they come to rest. The range of a particle of specific energy is well defined in a medium. In addition, they exhibit a Bragg peak, showing maximum energy deposits near the end of their journey. Several species of particle beams have been suggested and tried. The most common are proton beam facilities operating at 200–250 MeV. For a water medium, the ranges are 26 and 38 cm, respectively, for 200 and 250 MeV proton energies.[3] Figure 11.3 shows the energy deposits (dose) versus depth in a water medium for 250 MeV protons. Clearly seen is the Bragg peak at about 37–38 cm as anticipated. It is noteworthy that the peak is quite narrow, limited to about 2–3 cm. Thus it is effective in delivering a high dose to the tumor. At the accelerators, we can tune the beam energy to match the Bragg peak location to coincide with the tumor. Despite these advantages, proton therapy is much more expensive than gamma radiation therapy and it is not easily accessible. Thus the number of facilities is

[3]The website http://physics.nist.gov/PhysRefData/Star/Text/PSTAR.html is a resource to calculate stopping powers and ranges of protons in several media.

also limited. In the same vein, therapies with heavy ion beams such as carbon, boron, etc., are being tried. They are still in the experimental stage and are not commonly found.

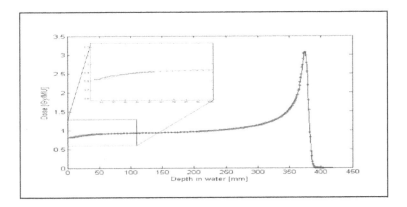

FIGURE 11.3: Energy loss distribution of 250 MeV protons in a water medium. The points are experimental data and the line is the theoretical calculation. The agreement between the theory and experiment for the energy loss and that of the range with the website is striking. (Adapted from Ulmer and Matsinos. European Physics Journal — Special Topics. Volume 190, Issue 1, 2011. arxiv.org/pdf/1008.3645.)

11.4 Trace Element Analyses

Trace element analyses find a wide variety of applications: archeology, art, environmental studies, national security, geology, nuclear science, etc. Nuclear techniques have been extensively employed, competing favorably with other methods in most cases. Ultimately, all nuclear methods rely on detecting and characterizing X-rays and gamma rays. Some of the techniques are real-time measurements as samples are exposed to particle beams, inducing atomic or nuclear excitations causing prompt emissions. A few methods rely on producing radioactive isotopes which will decay by emitting nuclear radiations with their characteristic lifetimes. The development of high resolution semiconductor

X-ray and gamma ray detectors revolutionized gamma spectroscopic methods of trace element analysis.

11.4.1 Resonance Fluorescence

The resonance fluorescence method is a real-time measurement of de-excitation of atomic or nuclear levels in response to exposure to the nearly white spectrum of X-rays and gamma rays. The basic physics principle of excitation of atomic and nuclear levels by these types of radiations is the same. The differences are only in the details.

If we shine a material with a photon beam of a broad energy spectrum, excitation of levels occurs for energies of photons matching the excitation energies of constituent atoms or nuclei. Most commonly, atomic excitations result in the emission of X-rays, and gamma rays are emitted in nuclear de-excitations.

While bremsstrahlung spectra extend from the X-ray region to much higher energies, modern day electron synchrotron light sources supply very high intensity X-ray beams. In recent years, light sources have become workhorses for material science by X-ray studies. Also, modern X-ray detectors offer excellent energy resolutions, permitting quantitative elemental analysis across the periodic table.

Nuclear resonance fluorescence is identical in principle to that of X-ray fluorescence, with a major difference being that it is sensitive to the nuclear structure of the isotope being studied. Thus, as in the case of neutron absorption, we cannot make a prediction of the sensitivity to the atomic number or mass number without resorting to experimental data. In general terms, nuclear resonance fluorescence involves photons of several hundred keV to a few MeV energies, which are highly penetrating radiation compared to X-rays. While X-ray fluorescence can probe the trace elements near surfaces, nuclear resonance fluorescence can reveal the materials deep inside. Thus there is growing interest in exploiting this method to detect the illegal transportation of dangerous goods in large containers by terrorist groups or other criminal organizations.

Several groups have been actively pursuing these investigations. Figure 11.4 is taken from Bertozzi et al., who demonstrated the sensitivity of nuclear resonance fluorescence to dirty bomb materials such as melamine $(C_3N_6H_2)$. Here one sees high energy gamma rays of

NRF Spectra: Water, Melamine and an Explosive Simulant

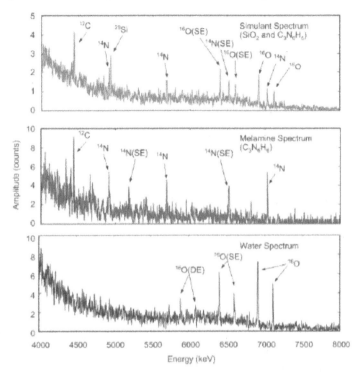

FIGURE 11.4: Nuclear resonance spectra of water (bottom) melamine (middle). The top spectrum is simulated with these elements and SiO_2 for gamma rays of 4–8 MeV energies. (Taken from W. Bertozzi et al. Nuclear Instruments and Methods in Physics Research Section B: Beam Interactions and Materials and Atoms. Volume 261, Issue 1-2, pages 331-336, August 2007. Elsevier.)

4.44 MeV energy from ^{12}C and 4.9 and 7.0 MeV gamma rays from ^{14}N isotopes. The 6.9 MeV gamma ray from an ^{16}O isotope in a water sample is also shown. In a given setting, we can measure the reference spectra for the intensity ratio of 4.9 and 7.0 MeV gamma rays from a melamine sample and compare them with the ratios of emissions from the sample under investigation. This should help us to avoid mistaken identifications.

Clearly, the attenuation coefficient of hard photons of a few MeV is much smaller than that of a few hundred keV X-rays. However, the cross sections of atomic phenomena responsible for X-ray emissions are several orders of magnitude higher than the nuclear processes responsible for isotope identifications. At this time, it is not clear if nuclear resonance fluorescence will prove to be a useful probe for security purposes.

11.4.2 Radioactivity Measurements

With the discovery of artificial transmutation in the early 20th century, nuclear science and technology revolutionized both the basic science and applications. Nuclear diagnostics by radioactive isotopes is just one example. Artificial transmutation is also a sensitive probe of trace elements. Neutron activation analysis, taking advantage of the high fluxes of neutrons at research reactors, is a well established technique, with sensitivities comparable to those of competing chemical methods with much less complexity. A neutron capture results in the excitation of a product nucleus to a high energy level. The product nucleus reaches the ground state by emission of characteristic high energy gamma rays which can serve, under favorable conditions, as a tool of trace element analysis. Also, often the ground state of a residual nucleus itself is radioactive, with characteristic emissions of a specific lifetime.

Currently, there are many tools for trace element analysis based on both nuclear and non-nuclear principles. While nuclear principles have their unique features, ultimately the goals of end-users, the ease of use, costs, etc., dictate the utility of a technique. In that sense, trace element analysis by nuclear techniques may be a thing of the past.

11.4.3 Non-Destructive Testing of the Strength of Materials

Several industries have a vested interest in the durability of materials under harsh conditions. An aviation industry may need to ensure the strength of an airplane's wing span. One may be concerned about the cracks in underground water supply systems which can contaminate the supply with pollutants by coming in contact with foreign materials. This can be an issue of public health. Small leaks in an underground oil pipeline might mean a huge loss of money besides possible hazards. Neutron radiography and gamma ray attenuation measurements are useful tools in this regard. It should be stressed that the principles of measurement and tomographic reconstruction are identical to those of computed tomography for medical applications.

11.4.4 Gamma Ray Attenuation Method

This is the same principle we have referred to many times in this book. Photons are attenuated as they pass through media. Variation in the compositions, especially defects in structures which create voids, will result in less attenuation or a high transmission of photons. We can thus scan the intensity profiles of transmitted radiation to create a to-mograph of a pipeline. One mounts a radioactive source such as ^{137}Cs or ^{60}Co and a detector, say an NaI(Tl) scintillator assembly, on a track, with the relative positions of the detector and the source fixed. The pur-pose is to measure the transmitted gamma rays for the source and the detector moving along the track to determine the transmitted intensities across the pipe's thickness. Any effective changes in thickness due to cracks in the pipes or some internal irregular structures will reflect in intensity variations. Just as we do for a SPECT imaging, we can con-struct images. The choice of photon energies should be optimized to get good image contrast. From among the isotopes with emissions of nearly the same energies, the economics, ease of handling and safety issues are the guide in making choices.

For example, the two workhorses of gamma ray measurements we encounter very often are ^{60}Co ($T_{1/2} = 5$ years) and ^{137}Cs ($T_{1/2} = 30$ years). Since ^{60}Co emits two gamma rays (1.17 and 1.33 MeV) com-pared to one gamma ray (0.661 MeV) for ^{137}Cs, the radiation levels are twice as high in ^{60}Co. In terms of laboratory use in tomography, ^{137}Cs is, relatively speaking, three times longer lived with respect to ^{60}Co.

As we note from Table 11.3, the mass attenuation coefficients of steel and air are not very different. However, the density of steel is 7.8 g/cm^3, while that of air is about 1.2 mg/cm^3. Thus steel is about 6000 times as effective as air in attenuating photons. The effectiveness of gamma ray tomography to identify voids in pipes due to cracks is clear.

TABLE 11.3: Mass attenuation coefficients [μ_ρ] of steel and air from XCOM. For ^{60}Co, μ_ρ(1.25 MeV photons) the average of 1.33 and 1.17 MeV emissions is given.

Gamma source	Steel (Iron = 0.98, Carbon = 0.02) μ_ρ [g/cm^2]	Air (Oxygen = 0.23, Nitrogen = 0.75, Argon = 0.03) μ_ρ [g/cm^2]
^{137}Cs (0.661 MeV)	0.073	0.077
^{60}Co (1.17 MeV, 1.33 MeV)	0.053	0.057

11.4.5 Neutron Radiography

Neutron radiography is an imaging system which relies on the distinct neutron interaction properties. For one thing, neutrons interact with atomic nuclei while photons interact with electrons. Thus photons are not very effective in imaging materials composed of light elements such as plastic, oil or water. Neutrons can penetrate heavy elements such as lead and they are sensitive to isotopes.

Neutron sources of a wide range of fluxes are available. A simple laboratory source of low neutron fluxes is composed of an alpha source and a beryllium (Be) foil. The alpha from a source such as ^{241}Am ($T_{1/2} = 432$ years) with a small Be foil will be a source of a constant supply of neutrons. They provide low fluxes of about 10^5–10^7 n/s. Particle accelerators provide higher fluxes of 10^7–10^{10} n/s. For higher fluxes, one accesses either neutron reactors or high energy, high intensity accelerators known as spallation neutron sources. Spallation neutron sources are usually proton accelerators of several GeV energy. They are used in conjunction with a mercury target. A nuclear reaction in this target results in very high intensity neutron beams. The neutron beams from high power reactors and spallation sources are comparable in intensities, varying from about 10^{11} to 10^{15} n/s.

As we have seen, low energy neutrons are more readily absorbed by materials. To this end, neutrons are slowed down in a moderating medium. After interactions or transmission through the materials under investigation, the outgoing neutron profiles are recorded in a detecting

Figure 1   Falcon no.54.547.

FIGURE 11.5: Photograph of the bronze sculpture of Egyptian Falcon in the Walters Art Gallery, Baltimore. [From Paul Jett, Sheley Sturman, Terry Drayman Weisser, *A Study of the Egyptian Bronze Falcon Figures in the Walters Art Gallery*, Studies in Conservation, Vol. 30, Number 3 (1985), pp. 112–118. www.maneypublishing.com/journals/sic and www.ingentaconnect.com/content/maney/sic.]

medium and data are analyzed. The principle of last stage is common to all experimental arrangements.

It may be most fitting to end this chapter with the images of an Egyptian falcon obtained by X-ray radiography and neutron radiography.[4] In Figure 11.5 we see the falcon figure in the display case of the gallery. Figure 11.6 shows radiographic pictures taken by X-rays and neutrons of an Egyptian bronze falcon figure in the Walters Art Gallery, Baltimore, USA. Hidden inside the bronze figure were the remains of a bird. The challenge was to know what was inside it without

---

[4]Paul Jett, Sheley Sturman, Terry Drayman Weisser, *A Study of the Egyptian Bronze Falcon Figures in the Walters Art Gallery*, Studies in Conservation, Vol. 30 (1985), pp. 112–118.

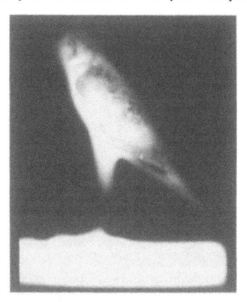

(a) X-ray radiograph of falcon

(b) Neutron radiograph of falcon

FIGURE 11.6: Radiographic impasse of the falcon in Figure 11.5. [From Paul Jett, Sheley Sturman, Terry Drayman Weisser, *A Study of the Egyptian Bronze Falcon Figures in the Walters Art Gallery*, Studies in Conservation, Vol. 30, Number 3 (1985), pp. 112–118. www.maneypublishing.com/journals/sic and www.ingentaconnect.com/content/maney/sic.]

destroying this historical artifact. Initially, a small hole was found in the head of the falcon figure. This led to an endoscopic examination, which revealed that there were some textiles and bones in the hollow figure. This led to further investigation.

Clearly, the display case picture is what we usually see with a naked eye. The X-ray radiography could not reveal the details of what lies within. Imaging with radiation from a 4 MeV electron accelerator was tried. The experiment did not yield any further details though the image was sharper. The next attempt was neutron radiography. Here the bones lying within the interior of the figure were clearly seen.

The species of nuclear radiations serve as tools for energy, science, industry, etc. While each type has its range of applicability, the usage of diverse techniques to supplement and complement each others yields a wealth of information.

A

Radioactive Decays

A.1 Radioactive Decay as a Random Walk Problem

Random walk problems have preoccupied physicists for a few centuries. As you all know well, a random walk problem appeared as the example of Brownian motion in fluids for which Einstein offered a solution. Prior to that, the kinetic theory of gases assumed that the molecules in a gas container are in incessant motion, undergoing elastic collisions with other molecules and walls of the container.

We may ask two different questions about the system and its evolution over time and/or space.

1. What is the probability that a molecule undergoes a collision in a time interval δt?

2. What is the average displacement of a molecule from an initial position and the variance of this parameter?

Let us consider the first question. Assume that the probability that a molecule collides with another molecule or wall does not depend on its past history, but it is simply proportional to δt, the duration of the time interval of observation.

Thus we have

$$q(\delta t) = \omega \times \delta t \qquad (A.1)$$

with $q(\delta t)$ as the probability that a collision occurs and ω the proportionality constant. The probability that a molecule avoids a collision during this time interval is

$$p(\delta t) = 1 - q(\delta t) = 1 - \omega \times \delta t \qquad (A.2)$$

For two successive intervals, $2\delta t$, we have

$$p(2\delta t) = (1 - \omega \times \delta t)^2 \qquad (A.3)$$

or, for n such intervals with $t = n\delta t$,

$$p(t) = p(n\delta t) = (1 - \omega\delta t)^n \tag{A.4}$$

$$\lim_{n\to\infty} p(t) = \lim_{n\to\infty} (1 - \omega\delta t)^n = \lim_{n\to\infty} \left(1 - \omega\frac{t}{n}\right)^n \tag{A.5}$$

or

$$p(t) = e^{-\omega t} \tag{A.6}$$

The exponential law is simply based on the assumptions

(i) that the interaction does not depend on the previous history and

(ii) n is very large.

We find the exponential law in several physical processes:

1. Radioactive decays

2. Discharging of a capacitor

3. Attenuation of radiation (light, neutrons, etc.)

A.2   Sequential Decays and Radioactive Equilibria

As seen in Chapter 2, a radioactive species decays via several channels, mainly sequential, sometimes through alternative modes of decay. The intermediate stages could be different chemical elements or they could be different excited levels of the same nuclear species or other combinations. Often, it is of interest to know the modes of decay, the rates of emissions of species of radiations, etc. The treatment below is quite general. We will begin with a sequential decay, involving three levels. We prescribe the initial conditions that a parent species of type A of decay constant λ_A decays to a daughter species of type B, which in turn decays to a stable granddaughter of type C. The decay constant of B is labeled λ_B.

The initial conditions are that at time $t = 0$

$$N_A = N_A(0) \tag{A.7}$$

$$N_B = N_C = 0 \tag{A.8}$$

Subsequently, the A species decays, feeding B and C at rates as follows:

$$\frac{dN_A}{dt} = -\lambda_A N_A \tag{A.9}$$

$$\frac{dN_B}{dt} = \lambda_A N_A - \lambda_B N_B \tag{A.10}$$

$$\frac{dN_C}{dt} = \lambda_B N_B \tag{A.11}$$

Equation A.9 simply represents the fact that the parent A decays with its decay constant, independent of the subsequent phenomenon. The second equation reflects that the daughter B is produced at a rate at which the parent A decays and it is depleted at the rate given by the product of its own decay constant and the number of B species present. The species C is produced at the rate at which B decays and the last equation is this statement.

At a time t, the number of A species is

$$N_A = N_A(0)e^{-\lambda_A t} \tag{A.12}$$

Thus

$$\frac{dN_B}{dt} = \lambda_A N_A(0)e^{\lambda_A t} - \lambda_B N_B \tag{A.13}$$

To solve for the activity and growth of species B and C, we need to do a little mathematical manipulation.

Multiply the above equation by $e^{\lambda_B t}$ and rearrange the terms to obtain

$$e^{\lambda_B t} dN_B + \lambda_B N_B e^{\lambda_B t} dt = \lambda_A N_A(0)e^{(\lambda_B - \lambda_A)t} dt \tag{A.14}$$

The left hand side of the equation can be recast as $\frac{d}{dt}\left[N_B e^{\lambda_B t}\right] dt$. Thus

$$\int \frac{d}{dt}\left[N_B e^{\lambda_B t}\right] dt = \lambda_A N_A(0) \int e^{(\lambda_B - \lambda_A)t} dt \tag{A.15}$$

Integration yields

$$N_B e^{\lambda_B t} = \frac{\lambda_A N_A(0)e^{(\lambda_B - \lambda_A)t}}{\lambda_B - \lambda_A} + D \tag{A.16}$$

where D is the integration constant which can be determined from the initial conditions as

$$D = \frac{-\lambda_A N_A(0)}{\lambda_B - \lambda_A} \tag{A.17}$$

to obtain

$$N_B e^{\lambda_B t} = \frac{\lambda_A N_A(0) \left[e^{(\lambda_B - \lambda_A)t} - 1 \right]}{\lambda_B - \lambda_A} \qquad (A.18)$$

The growth of the granddaughter C is, in magnitude, equal to the decay rate of the daughter B, i.e.,

$$\begin{aligned}
\frac{dN_C}{dt} &= \lambda_B N_B \\
&= \frac{\lambda_A \lambda_B N_A(0) \left[e^{(\lambda_B - \lambda_A)t} - 1 \right]}{\lambda_B - \lambda_A} e^{-\lambda_B t} \qquad (A.19) \\
&= \frac{\lambda_A \lambda_B N_A(0)}{\lambda_B - \lambda_A} \left[e^{-\lambda_A t} - e^{-\lambda_B t} \right]
\end{aligned}$$

We can distinguish two situations which depend on the relative magnitudes of λ_A and λ_B.

A.2.1   Case 1

The parent is very short-lived compared to the daughter $[\lambda_A \gg \lambda_B]$.

In this case $e^{-\lambda_A t} \to 0$ much faster than the term corresponding to the daughter's decay. Also, we can neglect the λ_B in the denominator. We then get

$$N_B = N_A(0) e^{-\lambda_B t} \qquad (A.20)$$

This indicates that the parent decays in a very short period of time. The number of daughter species is equal to the number of parent species in a very short time. Then the daughter decays with its own characteristic decay constant. Clearly, we are considering time scales which are much longer than $\tau_A \approx \frac{1}{\lambda_A}$.

Example A.1

^{241}Am decays with a half-life of 432.6 years as it populates ^{237}Np of $t_{\frac{1}{2}} = 2.14$ million years. In about 2200 years, more than 95% of the parent nuclei would have decayed to the daughter, which would then take nearly 8 million years to decay to 5% of its peak level.

A.2.2 Case 2

The daughter is short-lived compared to the parent $[\lambda_A < \lambda_B]$. Further subdivisions of this case are those when the daughter is very short-lived, $[\lambda_A \ll \lambda_B]$, or comparable to the parent lifetime $[\lambda_A \lesssim \lambda_B]$. We may rewrite the expression for $\lambda_B N_B$ as

$$\lambda_B N_B = \frac{\lambda_A \lambda_B N_A(0) e^{-\lambda_A t}}{\lambda_B - \lambda_A} \left[1 - e^{-(\lambda_B - \lambda_A)t} \right]$$

$$= \frac{\lambda_A \lambda_B N_A}{\lambda_B - \lambda_A} \left[1 - e^{-(\lambda_B - \lambda_A)t} \right] \tag{A.21}$$

Or we can write the ratio of the daughter/parent activities as

$$\frac{\lambda_B N_B}{\lambda_A N_A} = \frac{\lambda_B}{\lambda_B - \lambda_A} \left[1 - e^{-(\lambda_B - \lambda_A)t} \right] \tag{A.22}$$

For very large time t, the exponential term is zero and we have the ratio

$$\frac{\lambda_B N_B}{\lambda_A N_A} = \frac{\lambda_B}{\lambda_B - \lambda_A} = \text{constant} \geq 1 \tag{A.23}$$

If $[\lambda_A \approx \lambda_B]$, the daughter activity exceeds that of the parent and the daughter is said to be in transient equilibrium with the parent. Figure A.1 shows the plots of a parent and daughter activities for $\lambda_B/\lambda_A = 1.1$, normalized to that of the parent of time $t = 0$.

If $[\lambda_A \ll \lambda_B]$ (daughter very short-lived compared to parent) the ratio is nearly equal to one and the daughter follows the parent's half-life. The daughter is said to be in secular equilibrium with the parent.

In general, the number of atoms of the nth number in the sequence of decay with the parent only present at time t is given by

$$N_n(t) = \left(\prod_{i=1}^{n-1} \lambda_i \right) N_1(0) \sum_{j=1}^{n} \frac{e^{-\lambda_j t}}{\prod_{k \neq j, k < i} (\lambda_k - \lambda_j)} \tag{A.24}$$

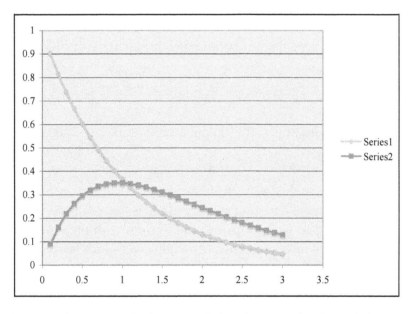

FIGURE A.1: Parent (Series 1) and daughter (Series 2) activity calculated for $\frac{\lambda_B}{\lambda_A} = 1.1$. The daughter is in transient equilibrium.

B

Energetics

B.1 Energetics of Nuclear/Particle Reactions

We present below the kinematics of nuclear or particle reactions where two bodies interact with each other to result in two or more bodies in the final state. We specify the bodies in the initial state as $a(m_a, E_a, \vec{p}_a, \vec{v}_a)$ and $b(m_b, E_b, \vec{p}_b, \vec{v}_b)$ in the laboratory, where m, E, $\vec{p}$ and $\vec{v}$ are, respectively, the mass, energy, momentum, and velocity of a particle. In the laboratory, the center of mass moves at a velocity

$$\vec{v_{cm}} = \frac{\vec{p}_a + \vec{p}_b}{E_a + E_b} \tag{B.1}$$

In the center of the mass frame, the center of mass is at rest and it is called the zero momentum frame. In this frame, the net momentum of the initial state bodies and also that of the final state bodies is zero. Thus, viewed from the center of mass frame, the particles in the initial state will approach with equal and opposite momenta. In the same frame, the two body final state products recede with equal and opposite momenta. In the many-body final state, the vector sum of the total momenta of all particles is zero.

In relativity, the total energy (E), mass (m) and momentum (p) are related as

$$E^2 = m^2 c^4 + p^2 c^2 \tag{B.2}$$

Thus the center of mass, a composite of the two bodies a and b has an invariant mass

$$M_{cm} c^2 = \left[(E_a + E_b)^2 - (\vec{p}_a + \vec{p}_b)^2 \right]^{1/2} \tag{B.3}$$

The above equation may be rewritten as

$$M_{cm} = \left[m_a^2 + m_b^2 + \frac{2}{c^4} E_a E_b \left(1 - \beta_a \beta_b \cos(\theta_{ab}) \right) \right]^{1/2} \tag{B.4}$$

where

$$E_a = m_a c^2 + T_a \quad ; \quad E_b = m_b c^2 + T_b \tag{B.5}$$

$$\beta_a = \frac{v_a}{c} \quad ; \quad \beta_b = \frac{v_b}{c} \tag{B.6}$$

$$\theta_{ab} = \theta_a - \theta_b \tag{B.7}$$

T_a and T_b are the kinetic energies of a and b, respectively. Here θ_{ab} is the angle between the a and b particles as they collide.

In low energy experiments, where we work with a stationary target, say b, and projectiles a, $v_b = 0$ and $E_b = m_b/c^2$, we have

$$M_{cm} = \left[m_a^2 + m_b^2 + \frac{2}{c^2} E_a m_b \right]^{1/2} \tag{B.8}$$

To produce particles c and d of masses m_c and m_d, we require $M_{cm} \geq m_c + m_d$; clearly the equal sign is the minimum, or threshold energy, for the reaction to proceed.

For the threshold of the reaction, we find

$$
\begin{aligned}
T_a^{min} &= E_a^{min} - m_a c^2 \tag{B.9}\\
&= \frac{M_{cm}^2 (min) - m_a^2 - m_b^2}{2 m_b} c^2 - m_a c^2 \\
&= \frac{(m_c + m_d)^2 - (m_a + m_b)^2 + 2 m_a m_b}{2 m_b} c^2 - m_a c^2 \\
&= \frac{-Q (m_a + m_b + m_c + m_d)}{2 m_b} c^2 \tag{B.10}
\end{aligned}
$$

where we used $Q = (m_a + m_b - m_c - m_d) c^2$.

At this stage, we approximate

$$(m_a + m_b + m_c + m_d) c^2 \approx 2 (m_a + m_b) c^2$$

Since, for most cases, $m_a + m_b \sim m_c + m_d \gg Q/c^2$, we may write,

$$T_a^{min} = \frac{|Q| (m_a + m_b)}{m_b} \tag{B.11}$$

A few observations are in order.

The M_{cm} is maximum when two particles undergo a head-on collision. The entire energy of the bodies goes to the mass of the center of mass, which is the maximum dynamical range that an experiment probes. So, instead of having projectiles hurled at a stationary target in the laboratory, it may prove to be useful to employ colliding beams.

If we have two identical particle beams of the same energy colliding head-on, then the total momentum of the system is zero. This is so because the momentum is a vector quantity. Two identical particles of the same energy moving in opposite directions have equal and opposite momenta. So the laboratory is a frame of zero momentum, i.e., the center of the mass frame. If we make the beams of particles and antiparticles of the same species and of the same energies, all properties except the energy are zero.

It is then possible to create several particle-antiparticle pairs of equal and opposite charges and neutral particles with net zero momentum subject to energy conservation. Several electron-positron and proton-antiproton colliders make extensive use of this principle to create new forms of matter. Examples are the Tevatron at the Fermi Laboratory near Chicago and the Large Electron-Positron (LEP) collider near Geneva, Switzerland. The LEP is now shut down to make space for the Large Hadron Collider.

Example B.1

An electron-positron collider with 50 GeV particle beams encountering head-on collisions is of 100 GeV center of mass energy. If one of the particles is at rest, for the 100 GeV center of mass energy, we have

$$M_{cm} = \left[m_a^2 + m_b^2 + 2\frac{E_a}{c^2}m_b \right]^{\frac{1}{2}}$$

Ignoring $m_a = m_b = 0.511 \times 10^{-3}$ GeV, we can still simplify to

$$M_{cm}^2 c^4 = 2E_a m_b c^2$$

or

$$E_a = \frac{100^2}{2 \times 0.511 \times 10^{-3}} \approx 10^7 \text{ GeV}$$

For a stationary target, we need to accelerate the projectile to 10^7 GeV. Energy economy of collider beams is evident.

C

Cross Sections

The terms cross section and luminosity are synonymously used to quantitatively indicate the probability that a nuclear or particle interaction occurs. The definition of cross section has its origin in classical physics. Second, unlike the conventional terminology where probability is simply a number ($0 \leq$ probability ≤ 1), a cross section is simply a measure of probability and it has dimensions of area (length square).

Definition of Cross Section

In classical physics, we notice that when two particles collide, the outcome is well defined if we know the direction and momentum of a projectile relative to the target particle. In a pool game, the directions and momenta of billiard balls are uniquely determined if the impact angle and speed of a striking ball relative to the stationary ball are specified. A good pool player takes advantage of physics to score well in a competition. The concept of cross section is an extension of this knowledge to the microscopic world. We immediately face a few complications. First, when we shoot particles or radiation, all we know is a beam of finite dimensions propagating in the laboratory; the smallest beam size we can determine is about a few microns, while particles themselves are nanometers (charged ions) or femtometers (protons) or even smaller. We make the stationary target of either solid materials or fluids (liquids or gases). Again, atoms and nuclei are invisibly small and we do not know which atom or nucleus is the partner in an interaction. Furthermore, unlike the contact interaction in a billiard ball scattering, the interaction is affected by fields, either electromagnetic fields of infinite range or nuclear fields of finite range. This means that projectiles do not necessarily undergo one instantaneous, local interaction with a target atom or nucleus, but they traverse through a field region over a small but finite amount of time. However, these regions and time in-

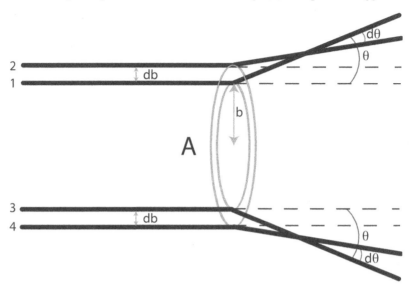

FIGURE C.1: Particles 1 and 3 approaching the target A at an impact parameter b are scattered at an angle θ. Particles 2 and 4 approach it at a corresponding value of $b+db$ and are scattered at an angle $\theta-d\theta$. Particles with impact parameters between b and $b+db$ emerge at angles between θ and $\theta-d\theta$. In the target plane perpendicular to the direction of propagation of particles, the spread of the impact parameters corresponds to an areal cross section $2\pi b db$. Here $2\pi b$ is the circumference of a circle of radius b.

tervals are negligibly small. For example, a positively charged particle sees no nuclear charge effect until it enters the atomic region since the target atom is of zero net charge. A charged ion of a few MeV, traveling at a speed of about one tenth the speed of light ($c = 3 \times 10^8$ m/s), spends less than a femtosecond (10^{-15} s) in the interior of an atom of dimensions of 1 nm or less. Thus we can assume that the projectile-target interaction is localized at a distance which we can call impact parameters, in analogy to the impact angle in a billiard ball game.

Figure C.1 shows the idealized geometrical diagram for a projectile approaching target A with impact parameter b. Note b is distance from the target and it has the dimension of length. The locus of the constant b is a circle of radius b drawn with the target atom A at its center and the direction of the projectile is perpendicular to the plane of the circle. From classical physics, we recognize that for each value of b, there is a specific deflection angle θ at which the projectile scatters. If there is a spread of db in the impact parameter with central value b, the projectile, after scattering, will appear within a corresponding angular spread $d\theta$, centered around θ. The locus of projectiles with impact parameters between b and $b + db$ is a ring of area $2\pi b\, db$ with target A at the center. Thus we see that all those projectiles appearing at angles between θ and $\theta + d\theta$ have passed through an areal cross section of $2\pi b\, db$. By counting the number of particles passing through specified angles, we can determine the cross section areas of target particles by which the projectiles are impacted. Thus the definition of cross section. This is the definition of a theorist connecting the probability of a deflected particle arriving at a specific angle to the impact parameter with a scattering center.

For practical applications and to compare with theoretical models, we have to provide an experimenter's definition of cross section. It uses the concept of solid angle.

Solid Angle

In an experimental arrangement, a beam of projectiles traverses a target medium and a detector measures the outgoing particles. Let us say it is at an angle θ with respect to the direction of projectiles. The detectors are usually finite sized. Depending on how far a detector is from the target-projectile interaction point, particles entering the detectors have an angular spread $d\theta$. Also, as they are of finite surface area, the lateral spread corresponds to an angle $d\phi$.

We can visualize it as in Figure C.2.

Here the detector of finite surface area dA is on the surface a sphere of radius R with the target at its center. For projectiles incident along the Z axis, the detector makes an angle θ with respect to the Z-axis and ϕ with respect to the X axis. It has angular spreads of $d\theta$ and $d\phi$ with respect to the Z and X axes, respectively. If we keep the detector at a fixed distance R from the center and rotate it around the Z axis, angle θ remains constant but angle ϕ changes as coordinates X and Y change. If we move the detector along Z while keeping both X and Y coordinates constant, the angle θ will change. The overall effect of the detector size and its location with respect to the target position can be

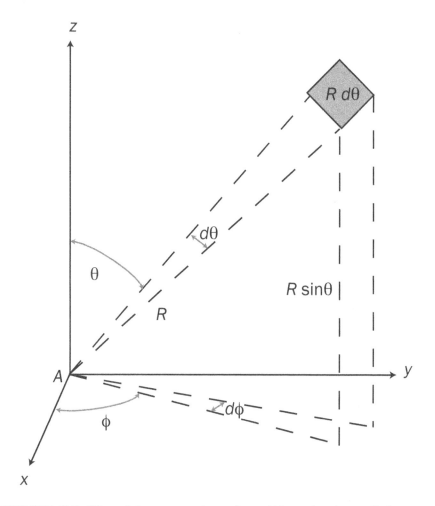

FIGURE C.2: Pictorial presentation of a solid angle. A small detector is set a distance R from the target point A. The detector subtends angles θ and ϕ with respect to the z and x axes, respectively. The corresponding angular spreads are $d\theta$ and $d\phi$.

expressed as a solid angle ($d\Omega$):

$$d\Omega = \sin\theta d\theta d\phi \qquad (C.1)$$

We can consider a detector arrangement in which it fully encloses the target so that all deflected particles enter the detector regardless of the angle of deflection. This corresponds to a spherical detector of radius R and the surface area of the detector is $4\pi R^2$. We get this result as integration over the angles θ and ϕ

$$\int d\Omega = R^2 \int_0^\pi \sin\theta d\theta \int_0^{2\pi} d\phi = 4\pi R^2 \qquad (C.2)$$

By convention, in such an arrangement, the detector is said to subtend a solid angle of 4π steradians.

Thus, for finite sized detectors located at a distance of R from the target, subtending an angle θ and extending over $d\theta$ and $d\phi$ polar and azimuthal angles, respectively, the solid angle is given by

$$d\Omega = \frac{a}{4\pi R^2} \text{ steradians} \qquad (C.3)$$

where R^2 is expressed in the same units as the area a.

Example C.1

A detector of area 1 cm^2 at a distance of 10 cm from the target has a solid angle

$$d\Omega = \frac{1}{4\pi \times 100} = 7.9 \times 10^{-4} \text{ sr}$$

The solid angle varies as the square of the distance. So, if we double the distance, the solid angle becomes $\frac{1}{4}$ of the initial value and so forth. Thus we are able to adjust the solid angle acceptance of the detector systems within the limits of possible configurations.

Now we are ready for the experimenters' definition of cross section.

Let us consider a target material of molecular weight M and density ρ. The number of target atoms per unit volume (N_{target}) is given by

$$N_{target} = \frac{N_A \rho}{M}$$

where N_A is Avogadro's number.

To proceed, let us recall the definition of flux (F). In physics, flux is defined as the rate of flow of a fluid, particles, radiation, energy, etc. In our context, it is the number of particles traversing the unit area of the target per unit time. Flux (F) has dimensions of area$^{-1} \times$ time^{-1}.

Clearly, larger fluxes with higher number of projectiles result in more interactions. Also, the number of interactions is proportional to the number of atoms/nuclei in the target. From the above reasoning, we can write that the number $N(\theta)$ of particles entering per unit time a detector of solid angle $d\Omega$ set an angle θ as

$$N(\theta) \propto F$$
$$\propto N_{tar}$$
$$\propto d\Omega$$

We can write these relations as equality with a proportionality constant $d\sigma/d\Omega$.

$$N(\theta) = \frac{d\sigma}{d\Omega} \times F \times N_{tar} \times d\Omega \tag{C.4}$$

The proportionality constant has the dimension of area, since flux is number of particles per unit time, per unit area. This is called the differential cross section, as it is specific to the angle of detection. The angle integrated differential cross section is called the cross section.

This definition of cross section in terms of flux may look a bit too restrictive as it depends on the beam dimensions. However, this is not the case. As long as the target is made big enough to ensure that all beam particles pass through it, the result does not depend on the beam size.

Say the beam is of cross sectional area a and N_{in} particles are incident on the target per unit time. Then, by definition,

$$F \times a = N_{in} \tag{C.5}$$

As these particles go through a target of thickness t, they see a target volume of at.

The number of target atoms intercepted by the beam is then

$$N_{tar} = \frac{N_0\rho}{A} \times a \times t \tag{C.6}$$

We then note that

$$N(\theta) = \frac{d\sigma}{d\Omega} \times \frac{N_{in}}{a} \times \frac{N_0\rho}{A} \times a \times t \times d\Omega$$

$$N(\theta) = \frac{d\sigma}{d\Omega} \times N_{in} \times \frac{N_0\rho t}{A} \times d\Omega$$

Thus we do not need to know the size of the projectile beam or the changes in their size, as long as the target is uniform and it is big enough to ensure that all particles pass through it.

Also, note that the term ρt, the product of the target density and its thickness, has the dimension of mass/area. As in the study of particle interactions, we note that this product is independent of the physical state (solid, liquid or gas) of the target material and is expressed in units of g/cm^2 or subunits.

We specified N_{in} as the number of incident particles passing per unit time and $N(\theta)$ as the intensity (number of particles per unit time) of outgoing particles. We may specify the total number of incident particles integrated over time, with N_{in} corresponding to the fluence of the particles. Then $N(\theta)$ is simply the number of particles.

In either case, it is easy to recognize that σ has the dimension of area, consistent with our earlier definition.

For charged particles as projectiles, one can measure the electric current or integrated electric charge as a measure of the total number of particles passing through a target.

For example, with either protons or electrons as projectiles, we know that each particle carries an electric charge of magnitude $q = 1.6 \times 10^{-19}$ C.

Therefore, 1 A of electric current = $\frac{10^{-19}}{1.6} = 0.625 \times 10^{18}$ particles/s. Particle accelerators put out beam currents as small as nanoamperes and as high as several amperes.

Example C.2

In a measurement, proton beams of 1 μA current were incident on an aluminum target of 1 mm thickness. A proton detector of 1 cm^2 area was set at a distance of 10 cm from the target. 30 protons/s are recorded in the detector. What is the differential cross section $(d\sigma/d\Omega)$ of scattering?

First, we calculate the number of target atoms:

$$\frac{N_A \rho t}{A} = \frac{6.023 \times 10^{23} \times 2.7 \times 0.1}{27} = 6.023 \times 10^{23} \, \frac{\text{atoms}}{\text{cm}^2}$$

Here we used Avogadro's number and $A = 27$, $\rho = 2.7$ g/cm^2 for aluminum of thickness $t = 0.1$ cm.

As 1 A of current is 0.625×10^{18} particles/s, 1 μA $= 0.625 \times 10^{12}$ particles/s.

The solid angle, as we see from the above example, is $d\Omega = 7.9 \times 10^{-4}$ sr.

The differential cross section is

$$\frac{d\sigma}{d\Omega} = \frac{N(\theta)}{N_{in}} = \frac{1}{\frac{N_0 \rho t}{A}} \frac{1}{d\Omega}$$

$$= \frac{30}{0.625 \times 10^{12} \times 6.023 \times 10^{23}} \times \frac{1}{7.9 \times 10^{-4}}$$

$$= 1.0 \times 10^{-31} \text{ cm}^2$$

The unit for the cross section is barn and it is defined as 1 barn $= 10^{-24}$ cm^2.

In this measurement, the cross section is
10^{-31} cm$^2 = 10^{-7}$ b $= 0.1\mu$b

As we can easily appreciate, one can measure cross sections from experiments by counting the number of beam particles and the number of particles in the detector of known geometry.

Here we assumed that each particle that enters a detector is recorded by the instrument. Often, physical instruments are not perfect machines and they do not necessarily record every particle. For one thing, we know that the interaction of radiation with matter is a statistical process, especially for neutral radiations. Thus some particles escape detection. We can define the efficiency of a detector system as

$$\text{Efficiency } \varepsilon = \frac{\text{number of particles detected}}{\text{number of particles entering the detector}} \tag{C.7}$$

Clearly, the efficiency is less than or equal to $1.\ 0 \leq \varepsilon \leq 1$.

In the above example, when 30 particles are detected, the number of particles entering the detector

$$\text{number of particles entering detector} = \frac{\text{number of particles detected}}{\varepsilon}$$

$$= \frac{30}{\varepsilon} \geq 30.$$

We must make efficiency corrections to determine cross sections. To this end, we rely on other sets of data for reference standards.

If we know cross sections either from the literature or from a theoretical prediction, we can plan to design experiments to achieve the desired counts.

Example C.3

The PET isotope FDG is produced by the reaction of proton beams interacting with an ^{18}O target to yield ^{18}F and neutrons. The ^{18}F is used to make FDG by electrochemical fluorination. The cross section for production of ^{18}F has a broad maximum of $\sigma = 0.4$ b for proton energies of about 8 MeV.

In commercial production using cyclotrons, proton beam currents of about 100 μA (6.25×10^{15} protons/s) are employed. If the target consists of water enriched in ^{18}O (instead of ^{16}O) we can estimate the number of ^{18}O atoms:

$$N_{^{18}O} = \frac{N_A \rho}{A} = \frac{6.023 \times 10^{23} \times 1}{20} = 3 \times 10^{22} \text{ atoms/cm}^3$$

where we used density $\rho = 1$ g/cm^3 and molecular weight 20 for water enriched in ^{18}O.

For a 1 cm long water column intercepting proton beams, the number of ^{18}O atoms $= 3 \times 10^{22}$ atoms/cm$^3 \times 1$ cm $= 3 \times 10^{22}$ atoms/cm^2.

number of ^{18}F atoms produced/s $= \sigma \times$ number of ^{18}O atoms/cm^2

$$\times \text{ number of protons/s}$$

$$= 0.4 \times 10^{-24} \text{ cm}^2 \times 3 \times 10^{22} \text{ cm}^{-2} \times 625 \times 10^{15}/\text{s}$$

$$= 7.5 \times 10^{13} \text{ s}^{-1}$$

$$= 2.7 \times 10^{17} \text{ h}^{-1}$$

Due to the finite lifetime, the growth of atoms is less than the rate of production. For the growth of a radioactive sample at a constant rate of production, we have (Section 2.4)

$$N(t) = R\tau \left[1 - e^{\left(\frac{t}{\tau}\right)} \right]$$

The half-life of ^{18}F is 109.77 minutes or 1.83 hours. The mean life of ^{18}F is $\tau = 1.443 \times t_{1/2} = 2.64$ hours.

$$N(t) = 2.7 \times 10^{17} \times 2.64 \times \left[1 - e^{\frac{t}{2.64}} \right]$$

after t hours of irradiation.

The activity is

$$A(t) = \frac{2.7 \times 10^{17} \times 2.64 \times \left[1 - e^{\frac{t}{2.64}} \right]}{3600} \text{ Bq}$$

$$= 7.5 \times 10^{13} \times 2.64 \times \left[1 - e^{\frac{t}{2.64}} \right] \text{ Bq}$$

It is easy to see that the maximum activity for this production is $A_{max} = 7.5 \times 10^{13}$ Bq, or about 2000 Ci.

It is of interest to note that after 1 hour of irradiation

$$A(t = 1 \text{ h}) = A_{max} \left[1 - e^{\frac{-1}{2.64}} \right] = 0.32 \times A_{max}$$

After 2 hours of irradiation

$$A(t = 2 \text{ h}) = A_{max} \left[1 - e^{\frac{-2}{2.64}} \right] = 0.53 \times A_{max}$$

Already, the second hour of production is about 65% as efficient as the first hour of production for this short lived isotope.

C.1 Luminosity

In experiments involving colliding beams, the term luminosity is used almost as synonym to cross section. We will define it here and provide examples.

Collider beams are of interest for at least two reasons. The first reason, as we mentioned earlier, is the high center of mass energies one can achieve in collider beam arrangements not possible with a particle beam incident on a stationary target.

Another advantage is the effective use of beam particles. In a stationary target experiment, beam particles pass through a target material once along their path. Some of them cause interactions and others are wasted as they are led to what are usually called beam dumps. In simple collider arrangements, two particle beams collide head-on in a region of interaction. In commonly used collider ring arrangements, these collision points can be arranged at more than one location, providing venues to run more than one experiment simultaneously.

In these arrangements, the event rate (N) for a physical process of interest is defined as below, when two particle beams 1 and 2 collide.

$$N = \text{cross section of the process} \times \text{luminosity} \qquad \text{(C.8)}$$

Event rate N = the number of events of physical process of interest per second.

The cross section for the physical process of interest = σ barns.

The cross sectional area of the beams crossing = a_1, a_2 for beams 1 and 2, respectively.

The number of particles in each beam crossing is n_1, n_2 in the two beams, respectively.

$$N = \frac{f \text{ collisions}}{s} \qquad \text{(C.9)}$$

For this arrangement, the event rate is

$$N = \sigma \times f \frac{n_1 n_2}{4\pi a_1 a_2} \qquad \text{(C.10)}$$

Thus the luminosity is

$$L = f \frac{n_1 n_2}{4\pi a_1 a_2} \qquad \text{(C.11)}$$

Luminosity has the dimension of the inverse of a cross section (σ^{-1}). It is usually expressed as nb^{-1}, pb^{-1} or fb^{-1}.

The luminosity of colliding beams is one inverse nanobarn (nb^{-1}) if the event rate is 1 count per second for a physical process of 1 nb (10^{-9} barns). Similarly, it is 1 pn^{-1} if the event rate is 1 count/s for a physical process of 1 pb (10^{-12} barns).

When feasible, this is achieved either by increasing n_1, n_2, the number of particles in each beam, or increasing the frequency, f, of collisions, or decreasing the areal size of the beams (a_1 and/or a_2) or a combination of some or all of these parameters. The technology and economics of experimental facilities play major roles in making the choices.

Instantaneous luminosity determines the count rates. One is often interested in the sensitivity of a measurement done over extended periods of time. We might ask how long it takes to get one event in the experiment for a physical process which is of 1 pb cross section. To this end, we specify integrated luminosity.

If the arrangement is of 1 nb^{-1} luminosity, it will take 1000 s of measurement to accumulate 1000 events for a 1 nb cross section, or 1 event for a 1 pb cross section. If we carry out such measurements for 1000 seconds, we achieve 1 pb sensitivity and the integrated luminosity is said to be 1 pb^{-1}. In this arrangement, the integrated luminosity becomes 1 fb^{-1} if the measurement is done over one million seconds (nearly 4 months of measurements).

Example C.4

In the year 2013, the ATLAS collaboration at the Large Hadron Collider of CERN published research results of their measurements of two photon yields for integrated luminosities 4.8 fb^{-1} and 20.7 fb^{-1} for 7 TeV and 8 TeV, respectively. This means that at 7 TeV, they see 4.8 events of two photon types for a cross section of 1 fb or equivalently one event for 0.21 fb.

The sensitivity was better at 8 TeV. At 8 TeV, they see 20.7 events for 1 fb or one event for 0.048 fb.

D

Physics of Semiconductor Detectors

Semiconductor detectors are, simply stated, solid state ionization chambers. They exploit the unique electrical conductivities of these materials. Generally speaking, we categorize materials as conductors, semiconductors and insulators based on their electrical properties of conductivity or resistivity. Conductivity (σ) and resistivity (ρ) are inversely proportional to each other. We might remember that the resistance of a piece of material is 1 Ω if a potential difference of 1 V across it results in a current of 1 A. A good conductor has low resistivity. The modern unit for conductance is Siemen (S) with the definition 1 S = $1/\Omega^{-1}$. The units for resistivity are $\Omega \cdot m$. The corresponding units for conductivity are S m^{-1} or $(\Omega \cdot m)^{-1}$.

The resistivity of a material varies with changes in external conditions such as temperature. Metals are good conductors at room temperature with resistivity of about 10^{-8} $\Omega \cdot m$. At room temperature, the resistivity of insulators is generally higher than 10^9 $\Omega \cdot m$. Semiconductors are those materials with resistivity anywhere between 10^{-6} and 10^3 $\Omega \cdot m$. Other materials are labeled either as poor conductors or poor insulators.

As the resistivity of materials varies with temperature, their electrical properties can be altered. As we know, some of them become superconductors of zero resistivity at very low temperatures. Metals (conductors) have the property that their resistivity increases with increasing temperatures. Generally, the resistivity of insulators decreases as temperature increases. Semiconductors have a somewhat intermediate property. At very low temperatures, they behave like insulators. Their resistivity decreases as temperature increases. At higher temperatures, the resistivity is nearly constant. At much higher temperatures, they behave like good conductors as the resistivity increases with increasing temperatures.

We can understand these features by asking what makes electric current flow in a material. It is the motion of charges in the medium and thus electric current depends on the mobility of charge carriers in a material. Good conductors are the materials where charges are freely wandering and a small electric field (volts/length) will set up the electric current. Insulators are the materials in which it takes a very strong electric field to make the same things happen and semiconductors are those for which moderate electric fields will do the job.

In physics terminology, we refer to this as the energy band gap. Just as we assign energy levels in molecules, atoms or nuclei, we can also assign energy levels in solids, which are a conglomeration of atoms or molecules. Instead of discrete, narrow energy levels for individual atoms, we call them bands of energy for solids as they are not discrete values. The outer most band of the highest energy is called the conduction band and the one just below it is called the valence band. Materials with a few electrons in conduction band are good conductors. Materials with no electrons in the conduction band are either insulators or semiconductors. If the gap between the conduction band and the valence band is large, requiring high energy (several electron volts), the material is an insulator. For semiconductor materials, a small energy supply will move electrons from the valence band to conduction bands.

A major breakthrough of 20th century technology was being able to manipulate electrical conductivities. It was realized that we can change resistivity by introducing foreign materials in an otherwise homogeneous medium in a controlled way.[1] Also, we can affect conductivities by operating them at different, controlled temperatures. This discovery led to several developments which revolutionized electronics, communications and several other industries. Radiation detectors also benefitted greatly from these developments.

Semiconductors are categorized into two types.

Intrinsic semiconductors They are pure materials. The band gap is finite, which renders them less conducting than metals. The band gaps of some of the most commonly used isotopes at room temperature are: silicon (Si) = 1.11 eV, germanium (Ge) = 0.66 eV and gallium arsenide (GaAs) = 1.43 eV.

---

[1]The present-day material science community calls this research area "band gap engineering."

These band gaps are smaller than those of insulators such as a diamond of 5.5 eV or silicon dioxide (SiO_2, sand) of 9 eV.

As the visible light spectrum spans about 2–4 eV, Si, Ge and GaAs will conduct if we shine light on them. It takes ultraviolet, vacuum ultraviolet radiation or low energy X-rays to induce conductivity in diamonds and sand.

Extrinsic semiconductors Here we introduce small amounts of foreign atoms into an otherwise pure, homogeneous medium. We categorize extrinsic semiconductors into n-type or p-type. If the impurity accepts an electron, it is a p-type semiconductor. It is an n-type if the impurity donates an electron. For Si and Ge belonging to Group IV of the Periodic Table, Group III impurities (for example, B, Al) make p-type and Group V impurities (for example, P, As) make n-type impurities.

Semiconductor detectors make use of both intrinsic and extrinsic semiconductors. The basic idea is to make a diode of the semiconductor. A conventional diode is said to be forward biased if the anode is at a higher voltage than the cathode. It is reverse biased if the opposite is the case, i.e., the cathode is at a higher voltage than the anode. When a diode is forward biased, electric current flows, and there is no current flow if it is reverse biased. Semiconductor detectors work on the same principle.

We make a junction of p-type and n-type semiconductors to form a diode. The p-type is like an anode and the n-type is like a cathode. Thus, if the p-type is at a higher potential than the n-type end, the diode is forward biased and current flows. This is not a useful configuration for a detector. We apply reverse bias, i.e., the p-type is at a negative voltage and the n-type is at a higher voltage. Then the positive charge carriers are held back on the p-side and the negative charge carriers are held back on the n-side. No current flows. The junction region acts like an insulator. This insulator region is called the depletion region because it is depleted of charge carriers. We can vary the depletion region by changing the voltage across the diode.

If a charged particle or gamma ray enters the depletion region and causes ionization, the negative charges move toward the p-end and positive charges go in the opposite direction. They set up current until all these charges are collected. The rest of the technique is identical to other detectors.

Semiconductor detectors are generally compact and provide very good position and energy resolutions. We can make strips of a few microns to achieve those position resolutions. The energy resolutions of semiconductors are a factor of 50 or so better than those of scintillation detectors. For charged particle detections, they offer nearly 100% detection efficiencies and very good energy resolutions. Quite often, these advantages of semiconductor detectors come with high costs and huge data acquisition hardware and software.

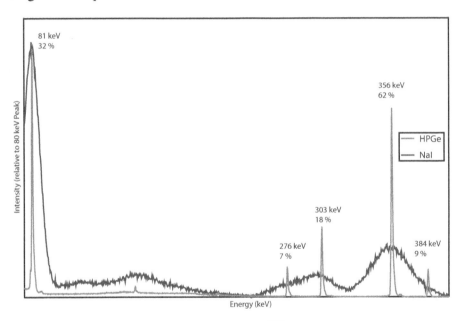

FIGURE D.1: Gamma ray spectrum of ^{133}Ba by an HPGe semiconductor detector, superimposed on a spectrum recored with a NaI(Tl) scintillation detector.

Figure D.1 shows a spectrum of ^{133}Ba recorded by an HPGe semiconductor detector and a NaI(Tl) scintillation detector. The energy calibrations of both detectors are matched to each other. The sharpness of full energy peaks in the HPGe detector is evident. While we see a broad hump consisting of 276 and 303 keV in the NaI(Tl) detector, they are cleanly separated in the HPGe detector. In this composite structure of the NaI(Tl) spectrum, we can still deduce that there are photons of two different energies (276 and 303 keV) as the width of the peak is nearly twice as large as that for a single line in this detector. However, for the complex of 356 keV and 384 keV, it is very difficult to see the 384 keV line in the NaI(Tl) detector, while they are well resolved in the HPGe detector. The advantages of the semiconductor detector for gamma energy spectroscopy is evident.

E

Websites

Here are some very useful websites for basic physics and nuclear radiations.

http://physics.nist.gov/cuu/index.html A good resource for fundamental physical constants and SI units. This website is an excellent resource for fundamental constants, definitions and values of SI units, their histories and conversion among units. It is updated every few years.

http://www.nndc.bnl.gov For a wealth of information about nuclear decays, reactions and radiations visit the National Nuclear Data Centre (Brookhaven, NY, USA) website. This website, along with its links, offers vast amount of resources for the public, students and researchers. You can find compiled and evaluated nuclear data of decays, reactions and structures, etc. There are several useful tools for scientists in basic and applied nuclear research. These databases are an invaluable treasure for everyone interested in nuclear science and technologies.

http://www.nndc.bnl.gov/wallet/ One can access an on-line version of the pocket book listing all isotopes, mass excesses, half-lives and decay modes. With this tool, Q-value calculations for nuclear processes are a simple task of addition and subtraction of a few numbers.

http://ie.lbl.gov/toi/ The Lawrence Berkeley National Laboratory runs this site. This is an on-line table of isotopes which allows for searches of radiations, atomic X-ray data, etc. It also lists all known isotopes of chemical elements and their decay modes.

http://pdg.lbl.gov The particle data group website and its mirror sites around the world are very useful resources for data, physics and instrumentation information. There are several useful reviews of instruments, propagation of radiation in media and useful physics and statistics to be found here.

http://www.iaea.org The International Atomic Energy Agency (IAEA) website is a good place to get a feeling for current activities of peaceful uses of nuclear radiations for health, energy, agriculture and industry. Upon subscribing to the website, you can access useful data.

http://physics.nist.gov/PhysRefData/Star/Text/PSTAR.html for proton stopping powers.

http://physics.nist.gov/PhysRefData/Star/Text/ESTAR.html for electron stopping powers.

http://physics.nist.gov/PhysRefData/Star/Text/aSTAR.html for alpha stopping powers.

http://www.nist.gov/pml/data/xcom/index.cfm For photon attenuation coefficient access. This website is a culmination of several decades of work of dedicated scientists. This provides us with the mass attenuation coefficients for elements, compounds and mixtures. It allows us to calculate the partial attenuation coefficients for coherent scattering, incoherent scattering, pair production and the photoelectric effect separately. Plots of data and downloadable tables are available.

http://www.icrp.org/ The International Commission on Radiological Protection (ICRP) is a not-for-profit organization with offices in London, England and Ottawa, Canada. Among other activities, they periodically publish recommendations of tissue weighting factors for different organs, which are of importance for radiological procedures.

http://www.world-nuclear.org/ The World Nuclear Association is an international organization dedicated to promoting nuclear energy, and it supports companies involved in the nuclear industry.

http://www.lightsources.org/ This website is a collaboration of communicators at light sources around the world. It is a good resource about current facilities and also those in the construction or planning stage.

F

Glossary

Atomic Mass Unit (amu) By definition the mass of ^{12}C is 12 amu. In this convention, 1 amu $= 1.66 \times 10^{-27}$ kg $= 931.5$ MeV/c^2.

Avogadro's Constant (N_A) This specifies the number of atoms or molecules in a one mole (kilo-mole) content of monatomic elements or polyatomic compounds. Numerically, $N_A = 6.023 \times 10^{23}$/mol. Here, mole is the weight in grams numerically equal to the molecular weight of the substance.

Boltzmann Constant (k_B) The kinetic theory of gases attributes the speeds and energies of molecules as responsible for all properties of gases and their influence such as the pressure they exert on container walls. According to this theory, the average kinetic energy of a molecule is proportional to the temperature of the gas and does not depend on the chemical nature of the molecule.

The Boltzmann constant has dimensions of energy $\times$ temperature^{-1}, numerically

$$k_B = 1.38065 \times 10^{-23} \text{ J} \cdot \text{K}^{-1}$$
$$= 8.62 \times 10^{-5} \text{ eV} \cdot \text{K}^{-1}$$

Brownian Motion This term refers to the random motion of molecules in a container in the equilibrium state. Robert Brown, a botanist of the 19th century, was the first to study this jig-jag motion. This is best observed by introducing a light foreign entity into an otherwise homogeneous medium, such as a light pollen into a water container.

Electron Volt A convenient unit of energy, commonly used in many subfields of physics. 1 eV $= 1.6 \times 10^{-19}$ J.

Higgs Boson The discovery of this particle was announced in 2012 after a decades long search by the particle physics community. Its mass is 125 GeV/c^2 = 222.5 × 10^{-27} kg. According to present-day theory, this boson is responsible for the masses of all other particles.

Lorentz Force A concise equation, known as the Lorentz force, provides a quantitative expression for the influence of electric and magnetic fields on a charged particle. We write the Lorentz force ($\vec{F}$) on a particle of electric charge q and mass m due to electric field ($\vec{E}$) and magnetic field ($\vec{B}$) as

$$\vec{F} = q\left[\vec{E} + \vec{v} \times \vec{B}\right] = m\vec{a}$$

It is important to note that for electric fields, the force is along the direction of the field. For a magnetic field, we find the force is proportional to the cross product (vector product) of the velocity ($\vec{v}$) and the magnetic field direction. Only charged particles in motion ($\vec{v} \neq 0$) experience force due to magnetic fields. For those in motion, the vector analysis tells us that the magnetic force is perpendicular to the directions of motion of the particle and the $\vec{B}$ field direction. A particle of mass m experiences acceleration $\vec{a}$ as above.

In the absence of a magnetic field ($\vec{B} = 0$), the particle experiences acceleration along the direction of the electric field. In the absence of an electric field ($E = 0$), the force is perpendicular to both the magnetic field ($\vec{B}$) and the instantaneous direction of motion ($\vec{v}$) of the particle. The particle is deflected from its path due to the influence of the field.

It follows a circular path of radius r, given by

$$F = qvB = \frac{Emv^2}{r}$$

where F, B, and v are magnitudes of the force, magnetic field and velocity of the particle, respectively.

Planck's Constant (h) In quantum mechanics, the energy of a photon is proportional to the frequency (ν) of the electromagnetic wave associated with it. Planck's constant (h) was defined to relate the frequency of a photon to its energy: $E = h\nu$ or

$$h = \frac{E}{\nu} = \frac{\text{energy}}{\text{frequency}} = \text{energy} \times \text{time}$$

Thus h can be expressed in units of energy (Joules) and time (seconds). Numerically

$$h = 6.626 \times 10^{-34} \text{ J} \cdot \text{s}$$
$$= 4.135 \times 10^{-15} \text{ eV} \cdot \text{s}$$

Quarks, Leptons, and Baryons Current understanding of particle physics is summarized in what is known as the Standard Model. Accordingly, quarks and leptons are two families of elementary particles. Currently, we know three generations of quarks and three generations of leptons. Electrons, muons, taus and their neutrinos are leptons. Free quarks do not exist but they are constituents of hadrons.

Baryons and mesons are hadrons, made up of three quarks and quark-antiquark pairs, respectively.

In this model, each quark is assigned a baryon number 1/3 and each lepton is assigned a lepton number one. The γ, W, Z and gluons are mediators of electromagnetic, weak and strong interactions, responsible for photon interactions, decay processes and binding of nuclei, etc.

Radiation Length In traveling a thickness of one radiation length of a medium, photon intensities and the power of high energy electrons are reduced by the factor of $1/e = 0.37$. Radiation length is the inverse of the linear attenuation coefficient.

Resonance Fluorescence When the excitation energy of a nucleus, an atom or a molecule is equal to the energy of an interacting photon, the photon is preferentially absorbed and the partner is in the excited state. The excited partner subsequently returns to its ground level emitting photons, characteristic of its level structure. This phenomenon is called resonance fluorescence. Resonance fluorescence of atoms or molecules occurs for very low energy photons up to about 200 keV energy. Nuclear fluorescence occurs from keV to several MeV energies.

Tomograph We encounter three terms: tomography, tomograph and tomogram.

- Tomography — a method of producing a three-dimensional image of the internal structures of a solid object (such as the human body or the earth) by the observation and recording of the differences in the effects on the passage of waves of energy impinging on those structures (*Merriam-Webster Dictionary*).

- Tomograph — the radiographic equipment used in tomography (*The American Heritage Medical Dictionary*).

- Tomogram — an image of a tissue section produced by tomography (*The American Heritage Medical Dictionary*).

We should note that a single tomogram (image) is rarely of much use for diagnostics. A comprehensive examination of a collection of tomograms for different cross sections of the target renders tomography a powerful tool.

Index